Reconstructing Wonder

CONTRIBUTIONS TO PHILOSOPHICAL THEOLOGY

Edited by Gijsbert van den Brink, Joshua R. Furnal and Marcel Sarot

Advisory Board:
David Brown
Vincent Brümmer
Oliver Crisp
Paul Helm
Werner Jeanrond
Christoph Schwöbel
Eleonore Stump
Alan Torrance
Nicholas Wolterstorff

VOLUME 14

Timothy Weatherstone

Reconstructing Wonder

Chemistry Informing a Natural Theology

Bibliographic Information published by the Deutsche Nationalbibliothek
The Deutsche Nationalbibliothek lists this publication in the Deutsche Nationalbibliografie; detailed bibliographic data is available in the internet at http://dnb.d-nb.de.

Library of Congress Cataloging-in-Publication Data
Names: Weatherstone, Timothy, author.
Title: Reconstructing wonder : chemistry informing a natural theology / Timothy Weatherstone
Description: New York : Peter Lang, 2017. | Series: Contributions to philosophical theology, ISSN 1433-643X ; Vol. 14 |
Includes bibliographical references and index.
Identifiers: LCCN 2017001544 | ISBN 9783631717530
Subjects: LCSH: Natural theology. | Philosophy and science. | Chemistry. | Aesthetics. | Religion and science. | Philosophical theology.
Classification: LCC BL183 .W43 2017 | DDC 210--dc23 LC record available at https://lccn.loc.gov/2017001544

ISSN 1433-643X
ISBN 978-3-631-71753-0 (Print)
E-ISBN 978-3-631-71754-7 (E-PDF)
E-ISBN 978-3-631-71755-4 (EPUB)
E-ISBN 978-3-631-71756-1 (MOBI)
DOI 10.3726/b10834

© Peter Lang GmbH
Internationaler Verlag der Wissenschaften
Frankfurt am Main 2017
All rights reserved.
Peter Lang Edition is an Imprint of Peter Lang GmbH.

Peter Lang – Frankfurt am Main · Bern · Bruxelles · New York · Oxford · Warszawa · Wien

All parts of this publication are protected by copyright. Any utilisation outside the strict limits of the copyright law, without the permission of the publisher, is forbidden and liable to prosecution. This applies in particular to reproductions, translations, microfilming, and storage and processing in electronic retrieval systems.

This publication has been peer reviewed.

www.peterlang.com

Acknowledgements

I must admit to being unsure when I embarked on this work as to how much I believed in natural theology: yes the world is often remarkable but what we do to it and each other is frequently not. Could God really be shown to be revealing of Himself through it all? And what of chemistry? I am forever fascinated by a whole host of facts, materials, gadgets and experiences, but could that be, can these be, of God? The results of these investigations I lay out below, but the power of the argument and the implications of the symmetries uncovered continue to amaze me. These are yet another example of God allowing us to view yet more layers and components of His extraordinary creative work.

E-mails are strange things: I sent some to Leiden University at my wife's urging in mid-2012 and Prof Drees very kindly gave me a direction in which to launch these studies. He also suggested someone who may be willing to supervise it. Another set of emails in early 2013 and Prof van den Brink pointed me towards Prof Marcel Sarot as one who might be willing to develop the proposal further. After many months of mutual exploration and more emails he graciously in the Spring of 2014 'took a punt' that it might be made to work, in-spite of my lack of knowledge in certain key areas, knowledge which he immediately set about improving. Prof Derkse then kindly came on-board to inform, validate and develop, both parts of the philosophy and importantly the chemistry. I am very grateful to these kind people.

I am indebted to Prof Jacob Claus of Saarland University and Professor Joachim Schummer of the HYLE journal, for most helpful suggestions around possible omissions and additions; to Mr Garth Cooper, a professional editor, for many comments on syntax and grammar; to both Canon Vernon White of Westminster Abbey and King's College London, and to Prof John Hedley Brooke, for illuminating conversations.

Having been given the opportunity to study, I sought permission from my superior, the Rt Rev'd Graham James, who was kind enough both to say 'yes' as well as to contribute financially. Thank you also to the leaders and people of the fifteen churches of the Barnham Broom and Upper Yare Group who voted to allow me to do this work and who have frequently had to suffer the results in innumerable sermons.

When many years ago I completed my MSc in Physical Organic Chemistry my then fiancée completed virtually all of the typographical layout from my inexpert diagrams. I failed to credit her for that work, much to my continuing shame. Now over thirty years of married life later, I wish to affirm her pivotal role also in enabling this present work to be completed.

Table of Contents

Introduction 9

Chapter 1: Religious Epistemology 13
    1.1 Introduction 13
    1.2 Contemporary Epistemological Approaches 15
    1.3 Conclusion 33

Chapter 2: Natural Theology 37
    2.1 Natural Theology within the Epistemological Framework: Introduction 37
    2.2 The Place and Relevance of Theologies of Nature 79
    2.3 Conclusion 81

Chapter 3: Chemistry and Natural Theology 85
    3.1 Appreciating Chemistry: the Historical Context and Contemporary Understandings 85
    3.2 Chemistry and Metaphysics 94
    3.3 Beauty as Bridge 118
    3.4 An Expanded Vision of Beauty 119
    3.5 Conclusion 123

Chapter 4: A Selective Survey of Current Organic Chemistry Research 127
    4.1 Purpose of the Survey 127
    4.2 The Selection Criteria 127
    4.3 The Papers 128

  4.4 Discussion 141

  4.5 Conclusion 142

Chapter 5: On God and Beauty 145

  5.1 Introduction 145

  5.2 The Two Streams 146

  5.3 The Aristotelian Stream 147

  5.4 The Platonic Stream 151

  5.5 Conclusion 155

Chapter 6: Discussion 157

  6.1 Introduction 157

  6.2 Defining Beauty 157

  6.3 Conclusion 162

Bibliography 165

Appendices 177

  Appendix A: Some Notes on Chemical Structures 177

  Appendix B: A Brief Introduction to Redox Reactions 180

Index of Names 183

Index of Subjects 185

Table of Figures 187

Introduction

The central argument of this book is that aspects of the natural science of chemistry as currently practiced, may inform a natural theology.

Firstly in chapter 1 I will seek to establish an epistemological methodology for the treatment of knowledge of and about the Christian God and of the justification of that knowledge consistent with contemporary understandings. This section on religious epistemology attempts to codify how the knowledge that follows later is justifiably rationally held. This is why it is present at all and moreover is why it is present at the start of this book. I believe we must lay the ground work of what can be rationally held before we can start to treat the material. Within this section on Epistemology the reader should note the prominence given to the tenets of a movement known as 'Reformed Epistemology' and within that, to the novel use of certain terms most notably 'justification'. These are explained there. It is perhaps unfortunate that two such well-known terms as 'reformed epistemology' and 'justification' should be re-used in ways which are quite different from their anecdotally 'obvious' explanations.

I will then in the second chapter review the current state of natural theology and seek to establish an approach within this discipline of systematics that is consistent with the epistemology established in the first chapter and that might build on an area of current chemical research. How should a natural theology, in the context of this book, be understood? The Gifford lectures, of which we shall have more to say below, deal with natural theology head on according to their founding principles. Their website by way of introduction, describes natural theology both as a classical discipline, and as a type of study in a contemporary nuanced form thus:

> A more modern view of natural theology suggests that reason does not so much seek to supply a proof for the existence of God as to provide a coherent form drawn from the insights of religion to pull together the best of human knowledge from all areas of human activity. In this understanding natural theology attempts to relate science, history, morality and the arts in an integrating vision of the place of humanity in the universe. This vision, an integrating activity of reason, is religious to the extent it refers to an encompassing reality that is transcendent in power and value. Natural theology is thus not a prelude to faith but a general worldview within which faith can have an intelligible place. (Gifford, 2016)

This book seeks to build upon this understanding of the function of a natural theology: no longer a 'proof' but a 'pulling together' of various insights, in this case from chemistry. It will do so in the context of the *conversation* that must inevitably take place when such a "general worldview" is promulgated. More specifically and particularly, this is a conversation between a researcher in chemistry who does not profess a Christian faith and a Christian natural theologian. Thus a working definition of a natural theology as stipulated here involves a presentation of rational inferences from knowledge gained by human activities to the actions of the Divine in creating and directing the Universe. Within this overall definition I go further in this book, in that by 'the Divine' I mean the Christian God, with the natural theology being presented by a Christian natural theologian.

I suppose this Christian natural theologian to be an 'orthodox Christian' by which I mean a person who treats the texts of the Bible, both the Hebrew scriptures and the explicitly Christian parts, as 'inspired by God and useful for teaching, for reproof, for correction and training in righteousness' (2 Timothy 3.16); who holds to what are termed the 'catholic creeds'; and who treats meeting and engaging in worship with other Christian people as part of his regular practice. I have explicitly not spoken of the degree of that regularity or the nature or type of that worship and neither have I made any mention of denomination. One might reasonably expect such an 'orthodox Christian' to be a regular attender at the place of worship of one of the mainstream Christian denominations, be that a church or a home or even a school. The reader will also notice that for the sake of brevity I have not defined precisely which books of the Bible should be regarded as canonical or part of it.

I imagine such a chemistry researcher, this partner in dialogue, to be an 'honest enquirer' who may for the sake of personal interest, investigate the claims of the person of Jesus Christ by visiting one or more meetings of one of these 'mainstream Christian denominations'. Wil Derkse speaks of the value of such a conversational approach when he remarks:

> The position of *dialogue,* or, as I prefer to call it, *conversation,* might be seen as a stage in a continuing process of integration of science and religion, both being human and cultural activities ..... *persons* active in separate domains can converse about contents, attitudes, evaluations, motivations (just a side-remark, perhaps *the* motivation in scientific practice is aesthetically), moral quality. (Derkse, 2001, p. 167)

Hence it will be plain that this book is to employ an attempt to hold in tension: *people, human beings,* who espouse a scientific approach to the elucidation of knowledge, together with those who also perceive matters in terms of Christian aesthetics.

Modern natural theologians often engage in somewhat of an obfuscation: as Chad Meister remarks, contemporary authors in this discipline do not suppose that their writings constitute proof of God's existence yet very much seek to demonstrate the rationality of theistic, not to say Christian, belief (Meister, 2013, p. 155). That said, such authors in their own personal conduct do not rest their own faith and practice on such a basis alone: they are frequently committed believers and practitioners and not uncommonly are ministers in their respective Christian denominations. There is therefore something of a subterfuge going on, or is there? Is the promulgator of a natural theology, whatever his proposed particular strategy, saying one thing but fervently believing another? Not at all, because natural theology is a common way of presenting the faith to 'honest enquirers' or at least to those who are not opposed to the Christian Gospel (in that such a theology is consonant with the faith of "ordinary believers", see Wynn 1999, p. 3). Such authors are offering an accessible methodology with which to approach the faith, to enable the Gospel to be seen as a rational and systematic response to contemporary life. In this present book I seek therefore as I have stated, to develop this argument as a *conversation,* as J V L Casserley has also suggested (Casserley, 1955, p. 7) between such an imagined honest enquirer, and in this case I imagine her/himself to be a chemist researcher, and myself offering the natural theological argument.

Among the contemporary physical sciences, chemistry stands out as one not generally thought to be helpful in contributing towards or informing, such a natural theology. Why this is so, is explained below. This project is therefore also an investigation into whether modern developments within the fields of epistemology, chemistry and theology might allow for a reappraisal of this hitherto largely accepted position. If this is possible, how might a proposed conversation of the type proposed above be re-invigorated, indeed re-legitimated, by and through contemporary revised understandings in epistemology? In chapter 3 I will discuss what is it about chemistry that lends itself to being implicated in such a revised appreciation. This will also require a brief historical survey of interactions between chemistry and theology from roughly the 17th century until the present time. In this chapter I will also expand on my opening remark above concerning the suitability of chemistry informing a natural theology.

Chapter 4 will include an analysis of a set of recent research papers in a particular area of chemical research. This analysis will look for commonalities in language, in assumptions and in methodologies so as to underline the common approach which I am proposing is used to speak in a cross-disciplinary manner.

This will be followed in chapter 5 by a brief thematic survey of the relationship between God and Beauty including the role of the researcher when framing his conclusions.

Finally in chapter 6 all of these elements will be drawn together to see what value might be obtained by applying this strategy to an appreciation of chemistry such that it may be allowed to have a place in informing natural theology.

The overall objective of the book is as has been said, to contribute towards the furthering of the dialogue between those in the disciplines of theology and chemistry: between Christian believers and chemist researchers wishing to enquire into the Faith.

Thus in short, the project will be addressed in the following order:

- The first chapter deals with epistemology and lays the groundwork for the approach presented in chapter two.
- The second deals with natural theology and within the wide definition given above, offers one that fits well with the epistemological approach presented in the first chapter.
- The third chapter describes chemistry in relation to its history, to metaphysics and to beauty.
- The fourth chapter surveys a small selection of contemporary chemistry research papers and comments on how they reinforce arguments rehearsed in the previous chapters.
- The fifth chapter gives a wider historical view on the subject of God and Beauty and crucially comments upon the role of the researcher's attitude to the Christian faith.
- Finally in chapter six I bring the preceding chapters together, offer a definition of beauty, check the definition's suitability and come to a conclusion.

Chapter 1: Religious Epistemology

1.1 Introduction

In this chapter my purpose is to answer such questions as: if I assert that I am justified in my belief in the Christian God and seek to assist others to also acquire this justified belief, how do I justify such a position of 'faith toward God' (Hebrews 6.1): what particular strategy or structure do I propose? By what means, using what methodologies do I justify my Christian belief; in short, how can I be assured that I know what I say I know? I start by providing an overview of contemporary approaches to religious epistemology. In so doing I hope to arrive at a considered personal position.

More generically Matthias Steup says of epistemology that:

> The debate over the structure of knowledge and justification is primarily one among those who hold that knowledge requires justification. From this point of view, the structure of knowledge derives from the structure of justification. (Steup,2013, p. 14)

In qualifying the justification of belief as needing to be 'rational' and implying also that it needs to be adequately structured, these authors are joined by Martin Smith who implies that the rationality of belief has also to do with the status of both the justification and the knowledge (Smith, 2014, p. 135); thus we need to be clear about what 'evidence' – the status of the justification -, and the assertion – the nature of the knowledge – mean for an individual protagonist in the debate. Such additional qualifications are a sign of the considerable degree of reflection that is current in religious epistemology. Peter Forrest defines 'evidentialism' to be 'the initially plausible position that a belief is justified only if "it is proportioned to the evidence"' and at times over the last two to three decades religious epistemology has swung towards strongly evidentialist approaches and then away again (Forrest, 2013, p. 1). Yet to state as Forrest does in the same place that: 'Contemporary epistemology of religion may conveniently be treated as a debate over whether evidentialism applies to religious beliefs, or whether we should instead adopt a more permissive epistemology' is to make the case too starkly: for one thing, it implies a uniformity to the idea of what 'evidence' amounts to, which as we shall see below, is itself doubtful.

It will become clear that not all the authors cited here are Christians themselves. In this way I hold that in order to facilitate the conversation proposed above, it is necessary to conduct the discussions in ways that are epistemically intelligible to those who are Christians and those who are not. Within contemporary epistemology various terms and understandings are used. For the purposes of the current book it is necessary for such a thing as 'knowledge' to be possible, about certain 'truths': that there is a God who was 'made known' by Jesus Christ (John 1.18), and that this knowledge can in some way be supported by or even revealed-in, the natural world around us. Current epistemology is a wide discipline and so for my current specific purpose I propose certain smaller categories or *theses* about justification and truth as suggested by Bruce Marshall, to enable the discussion to proceed (Marshall, 2000, p. 50):

- Christians may justify their position and beliefs on the basis of certain inner, in-the-mind, experiences; the sense of this is that such persons express their beliefs as interpreting that which they have experienced within themselves. We are not therefore speaking pejoratively of 'voices in the head' but rather of impressions and sensations which are then translated into narrative, or perhaps paintings or poems. Thus it is not the narrative or poem or painting which convinces the recipient of the truth of their belief, but the experience itself, and it is this latter which is committed to memory. See for example the account of the so-called Emmaus Road event in Luke 24: 'didn't our hearts burn within us' (Luke 24.32). This Marshall terms the *interiority thesis*.
- Christians are firm in their beliefs on account of what are to them certain self-evident items or pieces of 'incorrigible data' or beliefs which are logically derived from them. Such belief systems are described as being *foundationalist* and hence Marshall terms this the *foundationalist thesis*.
- Christians are able to explicate their beliefs in terms of criteria more universally held and that are as such not distinctively Christian. This Marshall calls this the *epistemic dependence* thesis.

Marshall further goes on to define in the same place:

- *a pragmatic* thesis, according to which Christian beliefs are justified by the communal and individual practices bound up with holding them true;
- *a correspondence* thesis, according to which the truth of beliefs, including Christian ones, consists in their agreement or correspondence with reality.

Thus according to these helpful categorisations and in recalling the desire to investigate specifically whether a natural theological approach might be most efficacious, there is already the hint that an epistemology that offers to the 'honest non-christian enquirer' into the Christian faith, an approach containing variously:

- some correspondence with reality
- arguments that are recognised by both Christians and non-Christians
- arguments that are held by a substantial and recognised community and are in that sense not of a single person only

might be the more promising of any group of approaches.

In what follows I discuss positions which themselves fall into one or more of the *theses* outlined above:

- (Wittgensteinan) Fideism
- Reformed Epistemology
- Prudential accounts
- Religious experience, memory and testimony
- Furthermore if the above represent more positive attempts to answer the question of the existence of God, we should also include some references to atheistic beliefs by way of balance. I end with a short conclusion summarising the chosen epistemological position in the light of the direction I wish to follow.

1.2 Contemporary Epistemological Approaches

1.2.1 (Wittgensteinan) Fideism

As Richard Amesbury explains, fideism is:

> the name given to that school of thought …. which answers that faith is in some sense independent of, if not outright adversarial toward, reason. In contrast to the more rationalistic tradition of natural theology, with its arguments for the existence of God, fideism holds …. that reason is unnecessary and inappropriate for the exercise and justification of religious belief. (Amesbury, 2014, p. 1)

As Stig Hansen remarks, when seeking to describe Wittgensteinan Fideism it should be distinguished from Wittgenstein's own thinking in philosophy of religion (Hansen, 2010, p. 1). In this sense Wittgensteinan Fideism functions as a label or name to a movement rather than an attribution of specific thoughts on Christianity to Wittgenstein himself, which were few in any case (Hansen, 2010, p. 1). None of this matters for our current study but it is useful to understand that the movement amounts to an application of Wittgensteinan principles of language by those who sought to promote his ideas. One such person was D Z Phillips and he described Wittgensteinan Fideism as an 'ill-conceived notion' (Phillips, 1993, p.xi) but as can be seen by careful reading, only because it was misunderstood. In contrast, far from being an advocate, Kai Nielsen describes Wittgensteinan Fideism as 'profoundly misguided' (Nielsen, 1973, p. 29). Thus creating an adequate definition of this movement is an exercise in coming to a negotiated view between advocates and critics.

In what follows I will provide a pair of quotations as attributed and then following this give a distillation of, or working description of, Wittgensteinan Fideism.

> …. Wittgenstein held the view that no belief at all should be considered as an underlying explanation of a given ritual. Rather, we should see ritual and religion as the natural expressions of a ceremonial animal. (Hansen, 2010, p. 3)

> The believer is not like someone who sees objects when they are not there, since his reaction to the absence of factual evidence is not at all like that of the man suffering from hallucinations….. When the positivist claims that there is no God because God cannot be located, the believer does not object on the grounds that the investigation has not been thorough enough, but on the grounds that the investigation fails to understand the grammar of what is being investigated – namely, the reality of God…. It makes as little sense to say, 'God's existence is not a fact' as it does to say, 'God's existence is a fact.' In saying that something either is or is not a fact, I am not describing the 'something' in question. To say that x is a fact is to say something about the grammar of x; it is to indicate what it would and would not be sensible to say or do in connection with it. (Phillips, 1993, p. 2)

On the strength of these quotations, Wittgensteinan Fideism amounts to a method of describing a perceived reality witnessed-to by a group of people, in which the specific language grammar (Wittgenstein uses the collective term 'language games' for such grammars) being used within that group of people, becomes the way of correctly understanding that perceived reality. In this way the words used by such a group of people mean what they do only in the context in which they are being lived-out.

Ludwig Wittgenstein himself in paragraph 7 of the *Philosophical Investigations*, speaks of 'language-games' being the means by which language is taught to children (Wittgenstein, 1958). On the function of such 'games', Wittgenstein gives the following in paragraph 499, amongst many examples:

> To say 'This combination of words makes no sense' excludes it from the sphere of language and thereby bounds the domain of language. But when one draws a boundary it may be for various kinds of reason. If I surround an area with a fence or a line or otherwise, the purpose may be to prevent someone from getting in or out; but it may also be part of a game and the players be supposed, say, to jump over the boundary; or it may shew where the property of one man ends and that of another begins; and so on. So if I draw a boundary line that is not yet to say what I am drawing it for. (Wittgenstein, 1958)

Given such talk of boundaries and fences, it might be supposed that those who hold to such a method of justification of religious belief, perhaps feel less of a need to interact with the world at large and Amesbury above certainly appears to be saying just that. Van den Brink characterises such a stand point as 'withdrawing into a bombproof bastion of fideistic starting-points about which no further rational discussion is allowed, and to which the rule of "take it or leave it" applies' (van den Brink, 2009, p. 176).

Yet there are two further characterisations of Wittgensteinan fideism which might be more positive: D Z Phillips shows that this method of justification may come hand-in-hand with a lively Christian faith. Phillips discards the idea that fideists have only a 'theoretical belief' and instead affirms them as being 'lovers' of God, not only lovers but in addition those who are passionate in their love (Phillips, 1992, pp. 90, 91).

A characteristic of Wittgensteinan fideism is that the words one might use to justify one's belief in God, will not make rational sense to those (non-Christians) listening. The reason for this is the language one uses is particular to one's own group and one should not expect to provide an intelligible rational justification to those who do not share one's faith (in this instance). Furthermore as Forrest implies, such a form of fideism 'is about attitudes not facts' and is 'only appropriate to Zen Buddhism and for some, relatively recent, liberal strands of Judaism and Christianity' (Forrest, 2013, p. 11).

Thus in conclusion on this point, fideism whilst potentially providing epistemic succour to the individual Christian, offers no rational explanation of the transformational Christian faith to those enquiring about faith in Christ. Since the Christian scriptures enjoin us to be able to express to others a 'reason for the hope that is in us' (an understanding taken from 1 Peter 3.15), it would seem that not to have any such reasons which might make sense to an enquirer, is a negation of that particular scriptural injunction.

1.2.2 Reformed Epistemology

The school of thinking known as Reformed Epistemology emerged in the latter part of last century with the philosopher Alvin Plantinga as one of its chief protagonists.

Before I outline the principles underpinning this significant new development in religious epistemology a word of warning about terms. 'Reformed Epistemology' is not merely an epistemological position espoused by thinkers of the protestant reformed branch of the Christian church: it amounts to a specific movement within religious epistemology. Plainly there remain people of the reformed protestant tradition who do not agree with the principles of 'Reformed Epistemology'. Furthermore in speaking of any beliefs that are 'basic', these should in the thinking of Reformed Epistemology be understood to be quite distinct from beliefs which are formally held to be 'properly basic'.

This distinction is important because the schema known as 'Foundationalism' made use of 'basic' beliefs to provide for a superstructure to justify belief systems. A natural theology where it seeks to provide evidence to justify rational belief in God presupposes a form of foundationalism. Douglas Geivett and Brendan Sweetman remark that it is because of the 'wide favor' previously enjoyed by foundationalism and its subsequent 'falling out of favor' that we are in this position now of exploring new epistemological positions, new ways of justifying religious belief such as Reformed Epistemology (Geivett and Sweetman, 1992, p. 4). Foundationalism as Marcel Sarot remarks:

> distinguishes between two classes of propositions: (1) Propositions of which the truth is clear without further argument: These can function as the foundation of our knowledge (basic propositions). (2) Propositions that need further argument to establish their truth …[thus] we can give full credence to a proposition only when it is a basic proposition, or when it is in a logically valid way derived from basic propositions. (Sarot, 2008, p. 260)

As Geivett and Sweetman remind us, natural theology was employed precisely to provide evidence for such basic propositions. It is in that sense an 'evidentialist' strategy. The work of Thomas Aquinas (1225–1274) as interpreted by some neo-thomists, is seen as being of great importance in this regard, since in his writings he provides such basic evidences or proofs of God's existence (the so-called Five Ways). It is upon such, as Vincent Brümmer points out, that a foundationalist epistemology including elements of natural theology, whilst plainly coming centuries later, could usefully be built (Brümmer, 1981, p. 208). Foundationalism as a term was coined at the beginning of the 1980s by Alvin Plantinga and Nicholas Wolterstorff 'for the deep-rooted assumption that they encountered far and wide in the history of Western thought, that the human mind can come to a real knowledge of the truth when we start from a solid 'rock-bottom', an indubitable point of departure' (van den Brink, 2009, p. 112).

The difficulties with foundationalism are multiple: 1) how might we know that a basic proposition is indeed basic and how can we be sure that our basic proposition itself does not come from another even more basic proposition? and 2) how might we justify the logical movement from one basic proposition to a derived proposition, this certain 'glue'? Furthermore should we accept point 1) just stated, we are in danger of continuing backwards ad infinitum, never being sure when we might encounter a basic proposition, one that we could rationally say was indeed basic. In addition as Brümmer points out, foundationalism (he uses the term 'rationalism') is too strict to allow for all that we might want to characterise as knowledge. Such knowledge might for example include characteristics of objects or events that are in addition to those that are directly observed and these he terms *impressive characteristics,* where one might be impressed with for example the beauty and wonder of such an event or object (these are my examples: see Brümmer, 1981, p. 210 where he mentions the sublime and mysterious).

Certain contemporary authors contrast 'foundationalism' with 'coherentism'. In this way for example Stanley Grenz and John Franke contend that:

> Coherentists [, therefore,] reject the foundationalist assumption that a justified set of beliefs necessarily comes in the form of an edifice resting on a base. In their estimation, the base/superstructure distinction is erroneous, for no beliefs are intrinsically basic and none are intrinsically superstructure. Instead beliefs are interdependent, each belief being supported by its connection to its neighbors and ultimately to the whole. Rather than picturing human knowledge as a building, coherentists draw from the image of a network in which beliefs come together to form an integrated belief system. (Grenz and Franke, 2007, p. 39)

Other authors again raise the profile of coherentism as an epistemically distinct position. Steup (2013, p. 18) goes so far as to say that coherentism is a competitor for foundationalism. He further makes the distinction (which I believe does have considerable merit) that whereas foundationalism consists of justified beliefs on the basis of either some form of epistemic privilege (which a Christian might infer to mean 'faith') or on the basis of some physical experience (which one might feel justified in saying is an accurate representation of the physical reality), a coherentist view would speak of your belief in for example the perception representing the reality. Thus Steup states that the foundationalist is contending that either of these two positions (that of epistemic privilege or experiential privilege) provide a methodology for establishing basicality. With that basicality the foundationalist may then construct their belief 'superstructure'. For Steup and as I stated above for Grenz and Franke too, the coherentist view is epistemically of an equivalent 'type' to foundationalism in that it provides a starting-point for justifying belief. Yet in my view this is unhelpful. Coherentism provides a methodology of marshalling propositions or components of a belief superstructure, but is unlikely to be able to provide the desired-for starting-point for justifying belief. It may have a use in this present book as a way of doing just that: of providing a *rationale* for the grouping together of certain strategies under one overall superstructure, but it is not a suitable place to start since the epistemological basis for belief is not being provided.

To return to Reformed Epistemology: this might also be understood as a schema or epistemic strategy for justifying Christian religious belief. It is important to understand that this strategy was not arrived at within a contextual vacuum. It is rather a strategy which is negotiated to sit firmly within the reformed position, for which to be true it must *a priori* uphold the notion of Christian faith being the supernatural gift of God 'of grace by faith' (Romans 5.2) and distinctly not something arrived at by an enquirer through logical deduction. The Apostle Paul expresses this further in addition to the words in Romans, in his saying 'For by grace you are saved through faith, and this is not from yourselves, it is the gift of God' (Ephesians 2.8). Since the process by which a person becomes a Christian is in that sense mysterious, being a supernatural gift of God, it follows that it cannot be based upon 'evidence' construed in the scientific understanding. It is thus not 'evidentialist' in contrast to Foundationalism and as Joseph Kim says, in an argument endorsed by Plantinga as 'careful, judicious and accurate', the 'Reformed Epistemologists, unlike Aquinas, reject the notion that one can offer a sound argument for the conclusion that God exists' (Kim, 2011, p. 3), although it must not be supposed that this movement rejects natural theology. Reformed Epistemology must however still rely on something being 'true' at the root of its system and yet the schema apparently does not permit evidence.

How is this dilemma to be resolved? A Christian person may look back upon their life and acknowledge that there was a time when they were not a believer and similarly that they are such now. Something must have happened. Plantinga and those who hold to his strategy need for this apparently supernatural event to be explicable in a way that is philosophically intelligible. Whatever happened cannot be susceptible to 'evidence' scientifically construed, but must nonetheless be rationally explicable. Reformed Epistemology's solution is to postulate that to the believer faith in God is self-evidentially true. It is as experientially true as water is wet or grass is green. In this way Deane-Peter Baker alludes to an 'immediacy' to one's knowledge of God (Baker 2007, p. 9). Such a fact is deemed to be 'properly basic' (Baker 2007, p. 8). There is no need, so this schema goes, for there to be any evidence for it. God is, and the believer knows this to be true. The Reformed Epistemologist sees no need to provide evidence for such an assertion.

In order to overcome the obvious defeaters such that for example a person might be deluded in their thinking something that to them is 'obviously true' and 'properly basic', Plantinga constructs a threefold platform of positive epistemic status' consisting of justification, rationality and warrant (Plantinga, 2007, p. 615). According to this platform, justification is achieved by the person being plainly sensible in holding their 'properly basic' proposition, where they could not be faulted in a moral sense for holding it to be true. (Let the reader note the alternate uses made of such terms as justify, justified and justification where for example earlier the quote used to define evidentialism as 'the initially plausible position that a belief is justified …..' uses this root in a different manner). Plantingan rationality is achieved where the beliefs are arrived at with an absence of cognitive dysfunction and may include beliefs held on the basis of the testimony of others and I would include scripture within such testimony. Warrant is somewhat different however and is that which translates a true belief into knowledge:

> The idea of our cognitive faculties functioning properly in the production and sustenance of belief is absolutely crucial to our conception of warrant; this idea is intimately connected with the idea of a design plan, a sort of blueprint specifying how properly functioning organs, powers, and faculties work. (Plantinga, 1993, p. vii)

In addition:

> The way to put it, then, is that a belief B has warrant for a person S if and only if B is produced by properly functioning faculties in an appropriate environment according to a design plan successfully aimed at truth. (Plantinga, 2008, p. 12)

Thus, as Richard Fumerton remarks, 'warrant' is that quality that converts belief into knowledge (Fumerton, 2006, p. 81). It is constitutive of whatever it takes to translate mere belief into knowledge and 'whatever it takes' could include reading the right materials, regularly practicing certain liturgies and actions, as well as being an active participant of a community of like-minded persons.

When Mark Wynn speaks of belief formation within a series of trust relationships, that are current within a given 'epistemic community', it appears reasonable to extend the understanding of such a 'community of like-minded persons' to being the Christian Church. Such beliefs are then justified within a web of both testimony and evidence, that also necessitates a practical dimension or outworking (Wynn, 1999, pp. 5, 120, 124, 125, 130). Wynn uses the same term 'properly basic' to signify beliefs held justifiably outside of the considerations of evidence, and yet as he himself says, his is a book setting forth a new form of the argument from design and as such within the same epistemic movement as natural theology (Wynn, 1999, p. 2). His 'evidence' is then generated through his arguments: broadly that the physical world shows forth such evidence, in the case of Wynn's specific argument, for the goodness of God. This latter is presented as a moral argument and so is not of interest here, yet the sense that Christian belief may be justified partly through a reliance on the collective noetic structure of the particular epistemic community I am part of, which I would call the Church, and must also contain an element of practical outworking, is significant. Thus Christian belief is something one does, as well as something one believes in, and indeed the significance of the two cannot be separated. It would appear from Wynn that we have translated a personally justified Christian faith into one that is relational and capable of being socially active through a series of trust relationships, or we might say from Christian tradition, relationships where we defer to one another in love (Romans 12.10, Ephesians 5.21).

This it would seem is one of the great advances in religious epistemology provided through the Reformed position, certainly since the 1980s: its reliance upon a property, this 'warrant', which translates Christian belief or some might say 'faith', from opinion into knowledge, by reason of the mechanism – namely the cognitive faculties – of the person or persons holding those beliefs. Thus warrant as stated bears an additive quality: it is a property which when added to Plantingan justification, gives rise to knowledge. Smith concurs that the Reformed Epistemological approach allows for warranted belief even when 'not supported by evidence' yet it seems to me that this might be more correctly stated as insufficient or inadequate evidence from the point of view of those opposed to the argument, since the Reformed Epistemologist plainly has justified or evidenced their belief(s) to themselves, and indeed Smith alludes to this later (Smith, 2014, pp. 137, 139).

Yet what of the other components of the Reformed Epistemological strategy? What of 'proper basicality':

> it should be relatively uncontroversial that many different sorts of beliefs can be properly basic. One main reason for this is just the fact that a wide variety of beliefs can be accepted, without cognitive dysfunction, on the basis of testimony, at least as long as the believer isn't aware of defeaters. The vast bulk of what I believe, I believe on the basis of testimony; the same, I dare say, goes for you. That I live in the US, that there is such a state as South Dakota-indeed, that my name is Alvin Plantinga – all of these things I believe on the basis of testimony. (Maps, birth certificates, histories of South Dakota are all, of course, forms of testimony.) Christian belief too, clearly enough, can be accepted on the basis of testimony without cognitive malfunction. Christian belief, therefore, can be basic with respect to rationality. (Plantinga, 2007, p. 615)

Plantinga then moves-on beyond this passage, to rehearse as he has done elsewhere, his *schema* for enabling him to assert that it is epistemically rational for Christians to hold certain Christian 'creedal beliefs' (Plantinga, 2007, p. 619). In this latter paper Plantinga is replying to an objector of his *schema*. It would seem that the result of Plantinga's *conversatio* is not a joining or a coming-together of minds, rather a more starkly delineated argument. I can and do admire his learning that allows me to justify to myself that I am not mad when I and my fellow congregants intone the Nicene Creed in the context of a Christian service, however I cannot agree that this is helpful or even inspiring for those who are merely visitors to such a service: the warranted justification appears to work for me and not for my conversation partner. Furthermore and perhaps crucially, given the above quotation, properly basic beliefs may be rationally held, in the Plantingan sense, by many who are not Christians. Indeed establishing 'proper basicality' speaks of a *methodology* which might well be effective for any number of sets of beliefs (see for example Plantinga, 2000, pp. 422–457 where he discusses religious pluralism). I have not in demonstrating proper basicality for my Christian belief-set, drawn anyone into a conversation with Christ. Thus it seems that we have reduced the great set of beliefs held by all humans, down to a smaller subset of those likely to be true, but have not pointed at the one that is true. The essence of a natural theology is that it is a rational, indeed more probable, explanation for an experience that is valid for and intelligible to, both believers and enquirers. It seems however that the intellectual exercise of demonstrating 'proper basicality' would be less useful in sustaining this conversation between such believers and enquirers.

In addition to the scriptural reasons given above, a further objection within Reformed Epistemology to the use, to the purpose, of any evidence for the existence of God lies in the theologically developed understanding of the effects of sin upon cognition, particularly as articulated within again reformed or protestant understandings of the Fall and of soteriology. As Baker explains, in the Calvinist schema, sin provides a degree of impediment to the non-believing (in Christianity) person, such that they are incapable of perceiving God (Baker, 2007, p. 7). Thus evidence, such as that provided by a natural theology, would be of no use to such a person as they are defined as being 'seeing yet not perceiving' (Isaiah 6.9, Mark 4.12). The non-Christian sinner then can only be susceptible to God's grace alone in finding Him, and not to evidence. The result is that it might be thought a Reformed Epistemological approach to justification differs little from the Wittgensteinian fideist approach:

> The Reformed tradition has insisted that the belief that God exists … may justifiably be found there in the foundation of our system of beliefs. In that sense, the Reformed tradition has been fideist, not evidentialist, in its impulse ….. Perhaps it would be just as well or better to point out to some inquirers that justifiably believing in God does not always require holding that belief on the basis of arguments. (Geivett and Sweetman, 1992, p. 149)

Yet Alvin Plantinga (1992, p. 134) remarks: 'many Reformed thinkers and theologians have rejected natural theology', although distinctively he does not include himself and the Reformed Epistemology movement amongst these, saying elsewhere:

> …. as he [Calvin] sees it, one needs no arguments to know that God exists. One who holds this view need not suppose that natural theology is of no use. In the first place, if there were good arguments for the existence of God, that would be a fact worth knowing in itself …. Second, natural theology could be useful in helping some-one move from unbelief to belief. (Plantinga, 1983, p. 73)

A distinct value in the Reformed position is that it firmly encompasses a realist approach, since this will in itself provide a useful bridge in the conversation between Christian and non-Christian chemical researcher. In this manner the character of 'religious' truth-value propositions is deemed to be the same as those of science and philosophy (Kim, 2011, p. 5). This is of great importance in the current project, for it allows us to move with ease from the method of justification (with 'justification' now being used in the more usually understood sense) to the substance of our enquiry confident that we remain intelligible to those who are not Christians, with this direction of thought taken from Alister McGrath (2001, p. xix). Thus it is now possible to distinguish the Reformed Epistemological approach from that of the fideist: if Christians are always to be 'ready to give an answer to anyone who asks about the hope you possess' (1 Peter 3.15), they must present an approach that is intelligible to those not of their persuasion. The Wittgensteinan fideist is able to say something like 'it would make sense to you if, whilst using language as we do, you also comprehended it from our point of view'; the Reformed Epistemologist in contrast is able to assert to the non-believer that they have a mutually intelligible justification of belief.

We have not stated how such a shared mechanism for agreeing that which is rationally held belief, might then give rise to the inference that the divine is at work in the natural world. Furthermore we are of course not asserting that mutual intelligibility between Christians and those studying the existence of God through an exploration of the natural world, equates to mutual agreement as to its origins. Overall however, the Reformed Epistemological thesis would appear to be very useful. It has been established that it is entirely rational to hold to the Christian faith. A natural theology is moreover a useful approach to be adopted towards someone 'moving' towards the Faith – precisely the use I wish to develop below. In reading Wynn it is possible to move from the perhaps overly person-centric view of epistemic justification proposed by Plantinga to those more community and relation-centric *loci* that are the expressions of churches existing within the framework of a Christian Faith founded upon the orthodox traditions. Finally as I have just mentioned, Reformed Epistemology encompasses a *realist* approach.

1.2.3 Interlude: On the Nature of Evidence

In this section I seek to show how a Christian may agree about the need for evidence for their beliefs, but that this evidence is of a different character to that as usually understood.

As I indicated above, Forrest's quoted view on evidence was too starkly drawn. For the proposition 'God exists' to be true from an evidentialist perspective, it is necessary to produce evidence that 'adequately supports' the contention (Smith, 2014, p. 139). Yet as Martin Smith says in answer to his own question 'adequate for what?': 'adequate for justification'. It is not clear what evidence could be given. What follows are certain observations on the nature of that evidence.

It is unlikely that a person, given the relative importance of say the existence of God, would rely on a single piece of evidence to justify their claim. Smith suggests that such a contention would be supported by cognitive processes of varying degrees of generality and indeed quality (Smith, 2014, p. 140). A believer is more likely to rely on a range of experiences, personal narratives and third-party testimonies. Thus in our present discussions of different epistemological approaches, it would be usual to expect any given person to offer not one, but possibly several elements of various strategies. Something of the quality of those 'elements of various strategies' will then inform the nature of the proposition being put forward. In this way, an understanding of the nature or quality of our proposition is based upon the nature of our evidence: a belief is supported on the basis of the evidence, something Martin Smith refers to as 'evidence basing' (Smith, 2014, pp. 140–141).

Wolfhart Pannenberg asserts 'Any intelligent attempt to talk about God – talk that is critically aware of its conditions and limitations – must begin and end with confession of the inconceivable majesty of God which transcends all our concepts' (Pannenberg, 1991, p. 337). In the Christian tradition and understanding it is the person of Jesus Christ who reveals, both historically as well as contemporaneously, this indescribable majesty. Thus the question of the justification of rational belief in God also begins and ends with the historical claims the Christian Gospel makes about the life, death and resurrection of the one man Jesus Christ.

When this consideration is then taken together with Smith's 'evidence basing', it is clear that we should expect something of the quality of that which we seek to prove, to be evident in the nature of the evidence we seek to prove it by. This is quite different from the nature of for example mathematical proofs, which are tautologous: in contrast, here the nature of the evidence is the proof we seek. The nature of the evidence that might point towards the existence of God, contains within it something of the nature of the divine: it is itself not the truth or the Divine we seek to portray, yet rather bears something of the quality of that which we seek to prove. We should therefore expect to see something of God in anything we might say points towards God. Again thinking back to the logical proofs in for example mathematics, the individual elements of an equation are not of themselves reflective of the answer (the quantity '2' is not reflective of the quantity '4' in the equation 2+2=4). In contrast Smith borrows the term 'evidentialist reliabilism' to describe a 'model of the world' in which a rational justification for belief in God might arise, for the 'religious believer' (Smith, 2014, p. 141). This evidential reliabilism is possibly an inelegant term yet it does helpfully combine the notion of warrant provided through cognition functioning correctly, combining it with a modified type of evidence of the sort we have described, producing the required-for evidence base. It might be countered that arguments of this type are in danger of shaping the divine to be merely a mirror of our own reflections and yet Smith is careful to point out that the evidence used relates to a divinity that is external to the person.

Thus, I reject the contention that there can be no evidence for the existence of the Christian God. Evidence that asserts the existence of God is only valid for the Christian believer and is of a different nature to evidence as understood for example in the mathematical sciences: it does in fact reflect something of the nature of the Christian God it points towards. Such evidence is in fact the work of faith, a 'gift of God' and so 'of God'. Evidence that could be effective in proving the existence of God to the non-believer, would amount to evidence that was in contrast independent of the God whose existence it sought to prove. If such evidence could exist it would then become necessary to qualify qualities attributed to God such as His omniscience or His all-powerful nature: there would be a sense in which such 'evidence' would be sitting in judgement over that which as Christians we say gave birth to it, namely God. This god could not be God. For this reason any such evidence for the existence of God, evidence that is independent of the God it seeks to substantiate, can not exist, and such would include that produced according to a scientific methodology.

1.2.4 Prudential Accounts of Justification

As Thomas Morris remarks, in being a 'simple, down-to-earth, practical, and decisive line of reasoning' (Morris, 1992, p. 257) to enable a rational belief in God, prudential justifications for theistic belief are, as Jeff Jordan remarks, a subset of pragmatic arguments (Jordan, 2013, p. 3). From this we might assume that these support Christian belief as well. Prudential beliefs 'are predicated upon one's preferences or goals or self-interest' (Jordan, 2013, p. 3). Thus a person might declare that they hold to such beliefs not because they necessarily believe them to be true, but on account of thinking that '*it would be best for me if I did*'. There is therefore something speculative about holding to such a belief system.

Let's remember that the goal of this opening chapter 1 of the project is to establish an epistemological 'platform' from which to explore a certain contribution towards natural theology, that might then – reasonably – direct an enquirer towards the Christian faith. Seen from such a perspective and again from Jordan (2013, p. 4), is it reasonable to enjoin our enquirer to agree with me, that it is rational to believe in God because of the probability of it being true, whether or not I can demonstrate the evidence (a truth-dependent argument) or indeed purely and simply because of the benefits this brings (a truth-independent argument)? Could such an approach be considered moral? Indeed James Cargile criticises such a prudential approach to justification precisely for this reason amongst others, that it is immoral (Cargile, 1992, pp. 283–289). Jordan though counters by reminding us that to not believe something that turns out to be more likely than not, would also be immoral (Jordan, 2013, p. 8).

On the subject of what God might think of the efforts of those who fake adherence to him, Cargile suggests that the Christian God if, as he says, He exists, is likely to be so 'nice' that he would not object to being treated in this manner. This of all his statements is by far the most interesting. It suggests that Cargile holds out the hope for himself that such is indeed the nature of the Christian God, presumably on the basis of examples Cargile himself has found in life since he does not report any investigation of Christian teaching on this subject (for example by reading the Christian scriptures in Isaiah 29.13, Matthew 15.8).

Cargile's remark is also perhaps the answer to one of the other objections to such prudential arguments: how could you know, if you are only 'being religious' by way of a bet or wager on yourself, which 'god' to pick [see note 28, William Lycan and George Schlesinger (1992, p. 270), see also Jordan, 2013, p. 6]? It seems that people have a notion of the character of this God they are enquiring about.

Thus the argument that Christian truth-claims might be possible and consequently would be worth further investigation, would appear to be a reasonable remark to make to someone enquiring about the Christian faith. Yet this would not amount to a prudential justification for such a person's entire 'faith-journey' throughout their life. Thus there exists the subtle distinction that I, in proposing a natural theology might 'sell' it to others in terms of 'try it out; it might be true' but would not use a similar argument to justify my own beliefs, nor would I seek such a justification as a long-term option for and in others.

1.2.5 Religious Experience, Memory and Testimony

Whilst discussing the Reformed approach above, I remarked that recourse to the warrant provided by a person thinking in a sound manner had much to commend it in terms of providing justification for belief. Yet this did not explicitly extend to experiential data but rather was directed at making belief in God a rational position to hold. An appeal by the Christian to a 'religious experience' brings in a wholly different and more problematic dimension. One might have what one terms a religious experience and one may or may not justify one's belief in God on the basis of that religious experience. There is also the issue that what one person perceives to be a religious experience, might not be acceptable to another.

In a section in his book, largely designed to refute any description of an event as being of God directly communicating with a human as true, John Shook helpfully lists all those ways in which one might be deluded into thinking that a 'mystical' experience, is actually God communicating supernaturally (Shook, 2010, pp. 99–110). These include what one might reasonably expect, including too much alcohol, an excess of nervous excitement, illusion, hallucination, errors in perception, lying etc etc. John Hick fortunately makes the assertion that 'we normally live on the basis of trust in the veridical character of our experience' (Hick, 1992, p. 307). Thus it is reasonable to evaluate sense data in terms of there actually being something in reality to assess, bearing in mind that such sense data can only be produced on the basis of the warrant provided through 'cognitive faculties functioning properly' (see above). This last would then again helpfully, remove most if not all of the items on Shook's list. Equally fortunately, Hick describes this understanding of our brain function as relying on 'trust' and things 'seeming' to be as they appear, since otherwise it would become impossible to adjudicate between competing accounts of the nature of theistic belief. Furthermore as Hick goes on to explain in the same place and mirroring Shook's account, in evaluating our experience we must be careful to filter-out perceived experiences that might have arisen through cognitive functions functioning *im*properly. We might usefully add to any such list, other factors which we know from personal experience have a tendency to alter our own individual judgement. In this way for example I know of several people who can do without sleep for prolonged periods and still operate effectively; I however do not share that skill.

Should such criteria be satisfied, Hick contends not only that it is entirely rational for us to be believed should we assert that we have had an experience of God, but moreover we can sensibly believe another who says such a thing. He uses Jesus as referred to in the Gospels as someone who may be trusted in this regard. Shook explicitly rejects this although only via a blanket statement that all scripture is rationally untrustworthy in regard to proving God's – or "a god's" as he says – existence. It would take only a single rational human to ascribe their belief in the existence of the Christian God as having arisen on the basis of scripture to refute that contention, and since there are many, Shook's point may be discarded

(Shook, 2010, p. 102). In summary, we can create potentially not only a web of contemporary acquaintances who have, whilst demonstrably acting rationally, experienced the presence of God, but moreover extend this web back in time to include others who have described experiences of a similar nature.

Shook (2010, pp. 99–100) disqualifies both groups as well as individuals who claim to have justified their belief in God on the basis of such 'religious experiences'. He does so largely on the basis that there are a great variety of religions who contend that their followers have experienced such phenomena. The underlying principle he presumably invokes, although does not state, is that not all can be correct or more specifically that only one can be correct at any one time. Yet this is itself difficult to accept: I see no reason why a Muslim can not have an intense experience of the love of God whilst not realising *per se* that it is Christ loving her/him, as a Christian would contend. Similarly, and most importantly for this project, surely almost all humans do or could express a great sense of wonder at the sight of a sunset or sunrise or some other great manifestation in nature e.g. the rainbow. It seems therefore perfectly reasonable to allow for individuals who belong to different religious groups to lend a supernatural explanation to the same event, without necessarily being aware of the identity of the 'god' who put it there, at least in the first instance. The Christian will offer an alternative explanation as to where such impulses arise from and with the warrant provided through historic precedent, attempt to explain the source, as seen for example in Genesis 41, Daniel 5 and Acts 8.26–40.

In an important re-telling and commentary upon one Christian person's reception of a 'religious experience', Baker offers a narrative originally recounted by Wolterstorff regarding the experience of a sudden powerful sense of being given a message from God (Baker, 2007, pp. 26–30). The person concerned acted upon the message as she believed she was directed to, which included relating part of it to her local church leader as well as other members of that church. Not only did she act as described, she also took herself off to a psychologist as she was concerned that she might be experiencing some form of mental illness. This professional recounted how such things were in the psychologist's experience quite common, and saw no need for the caller to be treated. With that it should be noted that with cognitive faculties having been assessed as functioning normally, a degree of cognitive Plantingan warrant had been achieved. Baker also states that apparently the person receiving the message was not of a Christian denomination where such things are commonplace. Interestingly Baker then goes on to relate the criticism Wolterstorff received, since he had decided – being of the Reformed Epistemological persuasion – that since the subject had correctly functioning cognitive faculties, the woman had indeed had a genuine communication from the Christian God. The substance of the criticism in the main, was that insufficient evidence and context had been provided to satisfy the contention that God had spoken to her, and that in simply applying the judgements he had, Wolterstorff was leaving the door wide open for persons with manifestly bizarre ideas, to also have similar epistemic justification for them. Yet what is not observed, is that in addition to the woman having proper warrant for her belief, quite plainly her group, her community, in fact her local Christian church, thought so too. There is therefore something here of the *pragmatic theses* of justification outlined above. Moreover the psychologist related that such experiences were perfectly normal and common. It is precisely this additional layer of epistemic justification – comprising testimony, community agreed memory and importantly *relationality* as providing justification – that would nullify those believing absurd contentions, since anecdotally we might reflect that for example, tales of communications from aliens are neither normal, nor common, nor likely to be agreed-upon by all members of a wider community (Baker, 2007, footnote to p. 30).

Similarly, Peter van Inwagen is apparently referring to the degree to which it is permissible to believe in the absurd (Van Inwagen, 2012, pp. 11–26). He is discussing how we can be sure that something is absurd in the face of no evidence, even though whatever the proposition is, *it is plainly absurd*. By way of example, he posits that there is an invisible (invisible because we do not have the technology to detect it even if present) teapot orbiting the sun between the earth and Mars. In his writing he appeals to two broad areas of general agreement: firstly that the means by which this came about cannot be reasonably and rationally imagined (he suggests aliens put it there) and secondly that we all know it is an absurd suggestion. As Baker implies above, we are appealing to a general sense in the population of what is sensible and what is not, and that the process by which the substance of the proposition came into being, is a valid point of justification of the knowledge of it being true or not. Thus in a Plantingan sense, the epistemic warrant for a proposition can be destroyed by a lack of general agreement within the group as well as a lack of an agreed pathway to achieve the proposition. Plainly not all propositions have pathways to their creation in this sense and yet here we are speaking of religious experiences, where one should reasonably expect such pathways to exist.

Thus in conclusion, Christian religious experiences, set within a context of the approval of a religious 'group' meaning the Church community, may be experienced by an individual functioning rationally, and when free from impediments as set out above, may reasonably be claimed to be genuine, as being from God Himself. Such attested experiences might then be used as arguments for the existence of God.

1.2.6 Objections to Religious (Christian) Belief

A particular type of objection to specifically Christian belief may be illustrated by the following Biblical parable which as commonly understood, is a narrative which is designed to illustrate examples of prevalent attitudes rather than recounting a single historical event. The specific pericope is somewhat lengthy and so by way of introduction it recounts how two men, one wealthy and another living in close proximity yet profoundly poor, both eventually die. The formerly wealthy individual, now resident in some unspecified form of hell, seeks to have messages sent to his family to warn them that their continuing evil deeds in the form of ignoring their poor fellow citizens, will land them in a similar predicament. The parable continues:

> But Abraham said, 'They have Moses and the prophets; they must respond to them.' Then the rich man said, 'No, father Abraham, but if someone from the dead goes to them, they will repent.' He replied to him, 'If they do not respond to Moses and the prophets, they will not be convinced even if someone rises from the dead.' (Luke 16.29–31)

This parable is illustrative of a more general scepticism:

> Whether religious experiences and testimony are able to provide adequate support for the existence of God will not [….] be something that can be settled just by reflecting on how strong these kinds of evidence seem – rather, it will depend on whether religious experiences and testimony are, in actual fact, reliably correlated with God's agency. A religious sceptic may deny that there is any such correlation but, for a religious believer, convictions about the origins of religious experiences and testimony will likely form a part of his overall worldview. The sceptic's charge that the evidence in question provides inadequate support for the existence of God will, then, be question begging – it will already take for granted a kind of non-religious worldview. (Smith, 2014, p. 141)

Thus arguments put forward in this book for the rational justification of religious belief should be seen to be aimed at the 'honest enquirer' into the Christian faith and not as either convincing proofs for the existence of God or the apologetic refutations of non-theistic commentators. Instead by portraying the Christian faith as a rational enterprise, it aims to draw others to investigate the claims of Jesus Christ.

Whilst therefore recognising the above, there are nonetheless certain strands to objections to Christian belief, which we might usefully survey now.

John Hedley Brooke outlines the major objections through a brief survey of those of Darwin. What is noteworthy (indeed it is the subject of his entry in the book) is how the objections to theistic belief come not from a positive affirmation of the truths of science yet rather from more general and indeed unscientific, observations of the apparent failure of organised religious practice. In this manner Brooke includes:

- the effect of preaching by Christian polemicists on the subject of hell and the likely destination of those who rejected the Christian gospel;
- the prevalence of pain and suffering in the world;
- the assertion of the existence of forms of divine revelation, given the apparent inconsistencies within and mutual incompatibilities of, various portions of the Christian scriptures;
- the inability to 'ascribe the contingencies affecting every human life to a designing and watchful Providence' (Brooke, 2010, p. 111)

Lesser issues that again were seen to speak against the existence of God included the degree of disagreement and even conflict between religious groups of the same persuasion; that others who were not religious could also in addition to Christian people, exhibit extreme moral rectitude and finally the indiscretions of leaders of a given Christian religious group.

Immediately, it will be apparent that there is no incompatibility between the holding of religious, indeed Christian beliefs and witnessing to the 'beauty and elegance' uncovered as a result of scientific research, with the result that the 'experiential and emotional aspects of the religious life' are not threatened by science (Brooke, 2010, p. 110). Brooke goes on to affirm that there are numbers of prominent scientists who are also fervent Christians. It might be reasonably expected that such highly intelligent persons had also appraised themselves of the above objections and resolved them to their own satisfaction.

Still others make various moral objections, which in large part focus on the less-than-exemplary conduct of Christian denominations and persons. The effect of such objections, as Donald MacKinnon tells us, is to force much-needed self-reflection upon Christians themselves rather than to generate fundamental objections to the faith (MacKinnon, 1963, pp. 11–34).

MacKinnon raises a further important matter:

> For Christians there is no escape from the issues raised by the involvement of the author and finisher of their faith in history. It is at once their glory and their insecurity that he is so involved. The very precariousness of our grasp of his ways reflects the depth at which he penetrated the stuff of human life. We cannot have that depth of identification on his part with our circumstances unless we pay the price of the kind of precariousness, belonging even to the substance of our faith, from which we may seek to run away to a spurious certainty even at the price of a kind of dishonesty which infects our whole outlook. We must be as sure as we can that we have rightly estimated the sort of certainty

which we can hope to have about Jesus and do not make the mistake of trying to make that certainty other than it is. (MacKinnon, 1963, pp. 31–32)

MacKinnon anchors the point of this present study accurately in the person of Jesus Christ. Furthermore in stating that there is 'no escape' he acknowledges the complete centrality of the identity and actions of Christ in history to its substance, in the 'depth he [Christ] penetrated the stuff of human life'. MacKinnon also goes some way towards negating the effect of any 'precariousness' causative of an 'insecurity', since it would appear difficult for something so apparently ephemeral to have such an essential effect.

1.3 Conclusion

Where have we arrived in our survey of contemporary epistemological approaches? What strategy would appear to be most appropriate in any conversation with the non-Christian physical scientist? Since 'faith is the gift of God' and God cannot be coerced into giving that same faith to any given individual, the focus of the Christian becomes that of providing a rational explanation for the epistemic position one holds as a practicing Christian believer, to any enquirer (1 Peter 3.15). This amounts to a conversation with such a presumed non-Christian. It must be held in language that is mutually intelligible, quite obviously. Consider such a conversation from the Reformed Epistemological perspective: it would amount to the Christian informing the enquirer that she was rationally justified in holding to her Christian beliefs and no more. From the Wittgensteinan fideist perspective it would amount to the Christian informing the enquirer that her Christian faith was internally coherent. Prudential accounts I would suggest offer nothing attractive at all in any ethically justifiable sense. We are left with offering reasonably imparted accounts of where the stuff of life and living demonstrates the enormity of the import of the Christian gospel; evidence that amounts to defeaters for those apparently overwhelmingly solid arguments for an entirely naturalistic explanation of all of what we can perceive.

And so like Smith I wish to 'tackle evidentialist objections to belief in God head-on' believing it entirely rational to hold to such beliefs when having the evidence for them (Smith, 2014, p. 145). Yet such a position would epistemically be only the starting-point, for there remain significant challenges once it has been accepted that belief in a god is rational:

- Can such a belief be rationally developed into belief in the Christian God; how may Christ be introduced?

- Quite how much justification is required or to put it another way, how much evidence is required to confer rationality?
- Orthodox Christianity includes truth-claims about Jesus Christ. It is plainly not sufficient to affirm 'belief in the Christian God' as a Christian: we must go beyond this and affirm our belief in who Jesus is.

On the approach developed above, this is now possible. The Christian believer, adopting the somewhat inelegantly-named mechanism of Smith's 'evidential reliabilism' may rationally sustain a belief in God. Yet I assert beyond this, that in belonging to and functioning as a part of a Christian tradition or 'church', being a group of liked-minded persons who also hold to and teach these same views and who affirm the orthodox ecumenical creeds, I may rationally hold to the scripturally expressed account of the life, death and resurrection of Jesus Christ, as penned by his followers after his ascension. These latter comprise the so-called New Testament Gospels, Acts, the Epistles and the Book of Revelation.

Moving on to considerations of evidence, we should expect the type of evidence offered for this rational belief to be reflective of the Object of our search. Thus we agree with Steup quoted above and attest that our epistemic structure – the typology of our knowledge – is reflected in the typology of our justification, of our evidence. In this way we would expect this evidence to conform to a pattern alluded to by the Apostle Paul when he instructed: 'brothers and sisters, whatever is true, whatever is worthy of respect, whatever is just, whatever is pure, whatever is lovely, whatever is commendable, if something is excellent or praiseworthy, think about these things' (Philippians 4.8). On a first assessment this looks like an Epistemology based not on reason yet rather on Faith – on revelation. Yet this is not what is meant. The typology of evidence may be contextual: for example a personal failure in a certain challenge may be construed as evidence for that person's lack of suitability for that given task or the very springboard required for future success. All that is being suggested here is that scepticism may blind the enquirer to good evidence, and that Paul's enjoining us to consider all that which is good, may enable a similar enquirer to value that which another had discarded. The Reformed Epistemologist says something similar: the cognitive faculties functioning correctly must be doing so in an 'environment in which it was designed to function' (Smith, 2014, p. 137). For our purposes here, such an 'environment' would include our conversation partner being able to view the evidence respectfully.

As for the level or quantity of required evidence, at first it might seem that for a belief as important as that of belief in the Christian God where I am required for example to daily take up my cross and follow Jesus (after Luke 9.23), the principle of pragmatic encroachment would apparently require a very high degree of proof (Smith, 2014, p. 144). Yet as Smith also shows in the same place, the practical considerations consequent upon following Christ – the perceived daily benefits – , whilst these do not amount to that which evidentialism requires, do act 'as a kind of catalyst, making it easier for one's evidence to [provide justification]'. To this principle then we do appeal in this project. In considering what this evidence might amount to, it will include but not be limited to, variously: personal, possibly infrequent, experiences duly affirmed by those in the church as being supernatural and sensations of the beauty of phenomena in the natural world, provided they conform aesthetically to the quotation from Philippians above.

It will have been noted that in constructing this project as in some sense a conversation between a Christian believer and an honest enquirer in the field of chemistry, there are self-evidently two parties involved in the discussion: the Christian listening and responding to questions and the 'honest enquirer' from the field of chemistry putting them.

Epistemically, Christians feel warranted in holding their beliefs on account of their properly functioning cognitive abilities within an environment which includes an active faith practiced in the context of a local expression of the Christian Church. The phrase 'active faith' implies a practice of the Christian faith which includes taking part in local and communal organised Christian services after the historic or orthodox pattern, as well as practical loving service towards one's neighbour after Christ's example, however imperfectly fulfilled. Possibly this believer will have had and/or continue to have some form of religious experience.

Chapter 2: Natural Theology

2.1 Natural Theology within the Epistemological Framework: Introduction

This book has as its aim the investigation of the possibility of 'a natural theology informed by chemistry', yet what does a natural theology entail? How should it be understood? James Barr tells us:

> Traditionally 'natural theology' has commonly meant something like this: that 'by nature', that is, just by being human beings, men and women have a certain degree of knowledge of God and awareness of him, or at least a capacity for such an awareness; and this knowledge or awareness exists anterior to the special revelation of God made through Jesus Christ, through the Church, through the Bible. Indeed, according to many traditional formulations of the matter, it is this pre-existing natural knowledge of God that makes it possible for humanity to receive the additional 'special' revelation. The two fit snugly together. People can understand Christ and his message, can feel themselves sinful and in need of salvation, because they already have this appreciation, dim as it may be, of God and of morality. The 'natural' knowledge of God, however dim, is an awareness of the true God, and provides a point of contact without which the special revelation would never be able to penetrate to people. Note that natural theology, thus understood, does not necessarily deny special revelation: it may, rather, make that special revelation correlative with a 'general' or 'natural' revelation that is available, or has been granted, to all humanity. But it does, in its commoner forms, imply that valid talk about God without any appeal whatever to special revelation is possible and indeed highly significant and important. (Barr, 1993, p. 1)

Useful though this understanding is, it does not tell us what a natural theology is for. It would seem by implication that it exists to enable 'valid talk about God', and yet only of a restricted form. There is a dividing line within it, for there is knowledge that may be known 'naturally' and knowledge that can only be known through a form of particular (divine) revelation. Natural theology is therefore a tool, a device to enable rational conversation between those who have already been enabled to know God through (a special divine gifting of) revelation, and those yet to do so, and who must therefore only know God 'naturally': it 'provides the Christian thinker with a point of contact or convergence with non-Christian thought which, from the apologetic point of view, may be of the greatest philosophical importance' (Casserley, 1955, p. xix). Barr also makes use of that same phrase 'point of contact' but goes further than Casserley in saying that without it that deeper or 'special' revelation of God to people is not possible. Also of interest is the way in which Barr states that natural theology does not deny special revelation and thus underscores the approach taken here and alluded to in the writings of Meister above: the Christian in promulgating a natural theology is not in any way 'cheating' in the argument. It is expected that they have an appreciation of the *destination* of their argument in that this is the acknowledgement of who Christ is. Equally such an approach as I am taking here still allows for an indeterminacy in the unfolding of outcomes, as I explore below.

From this we may agree with Rodney Holder that:

> It would seem that, for Christian faith to be commended in the modern world, natural theology is vital. (Holder, 2013, p. 131)

Once again Holder is underlining that which Barr says, that a natural theology is not merely a 'good to have', but essential if the Christian is to commend the Faith to our contemporary context.

We have spent some time attempting to navigate a way through epistemic research, developing a position that might enable a natural theology within the context of this project, to find a voice in the debate around the rationality of professing the existence of the Christian God. The question of rational justified belief in God was addressed through contemporary epistemological research; now the position of natural theology will be evaluated within this epistemic framework. Yet is there only one 'natural theology'? Russell Re Manning in the introduction to the recent *Handbook* from where the Holder quotation above was taken, says 'one of the primary aims of this *Handbook* [is] to highlight the rich diversity of approaches to, and definitions of, natural theology' (Re Manning, 2013, p. 1). He goes on in the same place to speak of 'new varieties of natural theology' and their 'complex diversity'. In this present book I propose just such a 'new variety'. We must therefore decide upon this (rational) natural theology which arises out of our chosen epistemological position such that we might then usefully dialogue with chemistry and chemists. In its widest sense such an approach to natural theology, as a tool enabling this rational dialogue within the constraints given, also validates its contemporary use, *irrespective of how it may have been seen historically.*

I will now proceed to outline some of the elements of such a natural theology.

2.1.1 The Audience

Chemistry as a discipline has exploded in terms of its significance over the last several decades as we shall discover below. It is practised by people in an arguably largely religiously neutral way in that no one is concerned to enquire of the religious persuasion of the university researcher in chemistry or the industrial worker in a factory producing agri-chemicals, for example. The language used by and between Chemists involved in research might anecdotally be expected to differ from that of natural theology. Immediately therefore a project of this type should be aware of its audience. Since I contend that the epistemic reasoning employed in Chapter 1 is there for both Christians and for non-Christians, and since the chemistry such as it is, may be read by all, this project is specifically not utilising a fideist approach, this meaning that the language being employed is that used in academic circles in Western Europe. This is of importance because even if the arguments employed in favour of a natural theology informed by chemistry are not convincing to non-Christians, they must be intelligible to them. Equally such arguments must be justified to and justifiable by, Christians, and this I have demonstrated in the chosen epistemology above.

Why does this matter? In his introduction to an important volume setting out a novel approach to natural theology Alister McGrath (2008, pp. 1–5) speaks of how his new approach is located in a newly articulated although not it must be said 'new', sense of the natural as informed by people transformed through their Christian faith. In this way he suggests that a natural theology is 'seen' through 'certain specific ways – ways that are not themselves necessarily mandated by nature itself' (McGrath, 2008, p. 3). He argues rightly I would suggest for a specifically Christian approach to natural theology. In so doing however he risks perhaps making this suggested new 'seeing' of natural theology opaque or at least unintelligible to the non-Christian, and in our case perhaps the enquiring non-Christian chemist. Such an approach if adopted in this book would miss our intended audience. He ends his extended essay (p. 315) with a quotation from the English polymath John Ruskin (1819–1900) which itself includes part of the following *pericope* from Ecclesiastes 3:

> What benefit can a worker gain from his toil? I have observed the burden that God has given to people to keep them occupied. God has made everything fit beautifully in its appropriate time, but he has also placed ignorance in the human heart so that people cannot discover what God has ordained, from the beginning to the end of their lives. (Ecclesiastes 3.9–11)

My purpose is not to enter into a debate about the translating of challenging passages of ancient Hebrew but merely to suggest that in its context this passage is speaking of the depth of wonder, complexity and elaboration of life that largely lies hidden from the human to the extent that (as the writer of Ecclesiastes then immediately goes on to suggest) all we should resolve to do is make the very best we can of our lives on earth. As McGrath himself says, this amounts to a theologically informed consideration of the subject of Ruskin's study – a consideration that would be immediately rejected by the non-theist as begging the question of the origins of both the capability of the (physical and metaphysical) sight as well as its object. The Christian would wish to maintain that our understanding of 'creation' is radically altered by and through the Incarnation – something which we know McGrath himself is keen to maintain (McGrath, 2001, p. 176). In summarising his discussion of 20th century developments, Holder acknowledges that this sense of confusion over who the audience is, for arguments from natural theology, persists: 'The fundamental ambivalence remains: are all these insights [from variously McGrath, Pannenberg, Barth, Torrance and others] simply confirmatory of beliefs already held or do they constitute arguments meant to command normative assent?' (Re Manning, 2013, p. 130). In this project at least it is I would suggest essential to maintain a language and a natural theology, that is intelligible to all of our audience, in order that Christian faith might continue to be 'commended' (see above) to the contemporary world. Such is therefore one aspect of this proposed natural theology: it is to be widely intelligible amongst the contemporary population including our conversation partner, the non-Christian chemistry researcher.

2.1.2 The Role of Order

I now digress to consider the issue of 'order' within the natural world, in the sense of how this word is apparently understood by various authors. This is of importance as certain natural theologies argue from the premise of the apparent order perceived in the natural world.

I have yet to discuss the modern natural science of chemistry in any detail, yet it will be clear from various introductory remarks, that those who pursue a natural theology have at times despaired of bringing chemistry into its orbit. The reasons for this are various but for our present purpose it will be sufficient to note that the study of mathematics and physics describes various set laws, apparently set into the fabric of the Universe (however so conceived). These are said to be immutable and as such are there to be discovered: we could for example call to mind the research being done at the CERN laboratories. Chemistry in contrast, since it makes new and transforms existing materials, plainly operates in a different way: it is utilitarian science or 'interventionist' and somewhat less 'contemplative' (Brooke and Cantor, 2000, p. 338). The practice of chemistry is also frequently messy and brings real physical skill into play. A sense of order might be discerned in the heavens or in the way fundamental particles make-up matter, but could this apply to chemistry?

Much is made of the words 'order' and 'orderliness' in McGrath's understanding of Creation. There are many examples in his *Scientific Theology volume 1: Nature* of which I note:

> Kepler's belief that there existed a fundamental congruence between the mind of God, human rationality and the fabric of the universe rests upon a classic insight of Christian theology, rigorously grounded in a Christian doctrine of creation. A scientific theology will wish to reclaim this neglected theme, and to affirm its importance, not merely for a right understanding of the relation between Christian theology and the experimental sciences, but for a proper grasp of the nature and scope of theology itself. (McGrath, 2001, p. 214)

And he notes the following conclusion from Brunner's work:

> Created human nature is such that it is able to discern the divine ordering of the universe. (McGrath, 2001, p. 204)

The difficulty here as Foster noted, is that such conclusions if misused could come dangerously close to suggesting that we are able to say *a priori* what God may or may not do. [As Bruce Russell explains, if I use the term 'a priori' of my (piece of knowledge) or proposition, I am saying that I hold something to be true, independent of experience (Russell, 2013, p. 1)]. Within this present project an 'a priori' understanding is something that might be worked-out from first principles, from presuppositions, from faith even and so known to be absolutely true without recourse to experiment or experience or any form of empiric investigation.] Such a mechanism implies that we may come to know God through reason alone. Thus in an analysis of the natural world as being ordered, what potentially can be lost in this understanding of the 'orderliness' of creation, is its contingency. What I am keen to establish is that a natural theology need not be only about a view of God that sees His creative acts as imposing order upon chaos; if God is being revealed in and through that which He has created, there is more to this portion of His self-revelation than Him saying 'I am an ordered God'! Yes we may legitimately speak of a certain dynamism within creation (McGrath, 2001, p. 288) but I would go further to suggest that Creation is contingent in the sense of 1 John 3.2: what we shall be has not yet been revealed. It is only in our relationship with God that we both know our eschatological destination and know that *we do not know what we shall become.*

Anecdotally, it is plain that both orderliness and disorderliness can be perceived in the natural world: a meadow of wild flowers is apparently disordered and yet it is only the framework, indeed a very carefully ordered framework, of natural conditions, which can possibly – in great vulnerability – give rise to the right conditions for the meadow to exist at all. There is an extravagant *dis*orderliness about a wild flower meadow: its profusion of colours and great variety of species giving often great pleasure to those who find it and succour to the animals that depend upon it. Our solution, to how a framework dedicated to orderliness can give rise to its self-same contingency, will provide the key to a comprehension of how a natural theology can with confidence understand if only in part, the Divine placing of order upon chaos in concert with its resulting complexities and disorder (Genesis 1.1–2). The reader will appreciate that such an understanding also reflects the tensions between perceptions of mathematics and physics as capable of upholding a natural theology and chemistry which is often seen as presenting difficulties in this regard. My present argument needs to move towards a theologically informed vision within a natural theology, that can resolve these tensions between ordered simplicity and the plain fact of nature's organic complexities yet all of this proceeding from some attempt at comprehending the God who made it.

What Foster, in a densely argued series of articles, so powerfully develops in his understanding of the intertwining of science and theology, is that an orthodox Christian theology allows for an 'undetermined-ness', a true contingency within both theology and the natural world:

> … a rationalist theology is logically bound to admit a further voluntarist element. The objects of God's reason, in so much as they are intelligible, must be universal, and no universal contains in itself a ground of the necessity of its own existence. It is not true merely that God need not resolve to materialise his ideas at all …….. God must therefore select from among his ideas those upon which he is to confer existence, and his act of selection cannot be necessitated by his reason, because there is nothing in any idea which would constitute it more suitable for existence than any other. (Foster, 1936, p. 19–20)

Thus from this, not only don't we know 'what we are to become' (1 John 3.2–3) but neither do we know why what *is* here, is here. There is nothing wholly set, nothing pre-determined in terms of what is to be created, at least such that we might come to have knowledge of it. Furthermore an understanding of any particular proposition may contain elements of justified knowledge that are known *a priori* as well as *a posteriori*. This is true *both of theological and scientific propositions*. Perhaps when Foster was writing this would have been a novel approach, certainly if made about scientific laws, yet this is now readily understood across the disciplines. Such a position is of importance to the current argument, since it allows for an active orthodox Christian faith which is worked-out on the basis of both things that are known, as well as things that may become known as the relationship between the person and God is developed, enlarged on, throughout life. From the perspective of the chemical sciences it is again of importance, as it allows for an holistic realisation of chemistry where elements of art and science, of experience and *a priori* Quantum-derived postulations, may epistemically validly be intertwined into a single synthetic understanding of a particular reaction or process. As I mentioned above, we have yet to understand chemistry in this way, yet already we have identified elements of a natural theology, carefully underpinned by epistemology, that allow for such a composition.

Yet Foster's work serves us further, possibly in even more substantial ways. For he is able to link the passion of love, of God's love shown in His decision to create, with a Christian rationalist theology. He declares that a rationalist theology should contain two markers (the second of which is quoted above, and the first here) of what he calls voluntarism:

> Although what God produces [may] be held to be completely determined by the ideas of his understanding .... His will must be arbitrary, in the sense of being free from determination by his reason .... The productive activity of the artificer differs from that of the lover, in that the former is guided by an object of reason, the latter by an object of desire. (Foster, 1936, p. 18)

Earlier Foster remarks:

> a divine Creator [.....] can embody his ideas in nature with the same perfection in which they are present to his intellect, so that the scientist can find in nature itself the intelligible objects of which he is in search, and not merely imperfect ectypes of them. (Foster, 1936, p. 15)

Hence we may readily understand that the scientist can expect to find 'intelligible objects' which are indicative of the God who is love (1 John 4.8), and who is in turn in love with His creation. In this way, as J B Stump remarks, the natural theologian acts in an *interpretive* manner towards her/his non-Christian colleagues in that the discovery of new truths about the natural order that might seem 'surprising and incredible' are no longer to be seen as incredible when viewed 'through the lens of Christian theism' (Stump, 2012, p. 148).

As so often in this writing, Foster's assertions overlap or appear to be contingent upon each other, as is here the case. It is exciting that Foster postulates that the exercise of God's free will is commensurate with a demonstration of his love. Furthermore to discern such markers in the natural world, the indeterminacy, the voluntarism, the combining of 'an empirical element [in these sciences] with their *a priori* character' (Foster, 1936, p. 18) is to see God's hand in what is perceived. It is in an abrogation of any 'right' to predict, that we might come upon God's self-revelation. And all of these come to us as markers of God's loving desire. True enough, we might only discern the truth of this by revelation and so move ourselves outside of the sphere of a natural theology. Yet the Christian acting as implied in 1 Peter 3.15 and offering 'an explanation for the hope that is in them', is able to offer such interpretations to their non-theistic colleagues within their natural theological conversation.

The natural theology proposed here will on account of the issues raised in this section, need to be in some manner showing-forth the love of God in creation. It will also provide an interpretive resolution to the tension between the orderliness that mathematics and physics tells us underpins the laws governing the natural world and the disorderliness made apparent in the contingency of outcomes.

2.1.3 The Role Played by Nature and the Natural

In considering the term 'natural theology', if we understand theology as being talk of and about God, then what of 'nature' and the 'natural'? McGrath (2008, pp. 3–5, 226) makes the very reasonable point that the Christian faith does shape a concept of the 'natural' and that a valid understanding of nature is 'a prerequisite for a natural theology which discloses the Christian God'. Yet such an approach does not guarantee a language that might be acceptable or even intelligible to the non-Christian enquirer as we have already remarked. Thus such a natural theology might well be helpful for the Christian community yet not for anyone else. McGrath goes on again to assert – and I would agree – that the Incarnation 'can be said to redeem the category of the 'natural' allowing it to be seen in a new way' (McGrath, 2008, p. 4). Once again though, such 'seeing' might be challenging for those who do not acknowledge Christ.

In our discussion of prudential accounts of epistemic justification, it was noted that some authors had suggested that following a religious faith was rational on the basis that to do so was a sensible bet, given the uncertainties of scientific discernment. If I were to suggest that a person should at least investigate the claims of the Christian faith on the basis that it is rational and offer if not a convincing proof of God's existence at least the sense that it is 'not improbable', some might think I was weakening my claim to have knowledge of God's existence, of warrant for the belief that God exists. It might be thought by a the non-Christian that my postulated natural theology was providing reasonable grounds for belief and not proof of knowledge. In turn, it might be thought by a Christian apologist that I was suggesting the Christian faith could be adopted solely on the basis of reason: that a person might arrive at the conclusion the Christian faith was entirely true, having followed a rationally reasoned argument through a series of steps. McGrath in the above-mentioned volume is to be commended for pointing out that the Christian faith involves a transformation of the person through an encounter with the actual person of God, initiated not on the basis of for example, that person's character or intrinsic goodness or suitability, or indeed for any other reason (including Reason itself) than that God decided to do so (McGrath, 2008, pp. 4–5). There is therefore an epistemic gap between a knowledge that says 'there is a God', and a human person who says 'I am a Christian'. Scripture witnesses to such a gap 'You believe that God is one; well and good. Even the demons believe that – and tremble with fear' (James 2.19). To have been enabled to bridge this gap is to have experienced God through His grace by revelation, which a natural theology cannot speak of since it evinces revelation. The objective of my natural theology therefore is not to prove that God exists, since such proof (through Reason alone) does not yield the transformation through God's grace, of the person in the face of an encounter with Christ. The purpose of a natural theology is to lend rationality to the thought that there is a Creator of the Universe and in the manner presented here, a Universe redeemed through Christ. This theology is presented in a language that speaks rationally (and evidentially) to those not (yet) of that persuasion. It gives the hope of a reality beyond Charles Gore's 'faint and flickering gleam' (Gore, 1891, p. 77). This is in part then the 'new' aspect of the natural theology being proposed: not a proof of God's existence but an aid in our conversation to point towards the Christian God.

Can we identify such a natural theology and in what way might it be thought 'natural'?

Re Manning in his introduction to the major new *Handbook* to the discipline quoted from above, can provide 'no easy answer' to the question of what natural theology is and goes on to speak of the diversity of approaches and definitions of natural theology (2013, p. 1). For the purposes of this project it is therefore necessary for the reasons already set out, to define which particular 'natural theology' we are speaking of.

I will be speaking about the role of chemistry below, yet immediately in relation to the 'natural' of a natural theology, any such theology that only relates to pre-existing structures and laws, as elucidated through for example mathematics or physics, would be insufficient to involve chemistry simply because this particular discipline involves the transformation of material from one form into another. Frequently it also involves the creation of entirely new compounds or materials. Anecdotally it appears that many find describing such materials as being 'natural' problematic: in what way for example could modern polymers or plastics be considered 'natural'? Given that less and less of the earth's surface is unaffected by human activity I believe that it is timely to reconsider what it is that we consider to be 'natural'. This is the conceptual bridge that must be crossed for chemistry to be considered in any way 'natural'.

I would also not wish to appeal to a natural theology reliant upon any notion of a *sensus divinitatis* as being that inbuilt perception of God inherent and inherited by all humans, as spoken of by many commentators, not least John Calvin, yet see for example Justin Barrett (2012, p. 324). I would therefore, using the word 'natural' in a different context, reject the idea that there is anything natural about humans simply knowing that there is a God. I would do so not because I disagree with the Apostle Paul that such exists: 'because what can be known about God is plain to them, because God has made it plain to them' (Romans 1.19) but rather precisely because humans, again as Paul says in the same place, are so very adept at ignoring it. As Paul points out in the preceding verse, people 'suppress the truth in unrighteousness'.

Taking these points together I would therefore wish to encourage a revision of our perception of the 'natural' world, to include that which nature itself, on account of its laws, *permits* to be created, in addition to that which apparently came about of its own (that is nature's) volition. Furthermore any appeal to Romans 1 as espousing a natural theology is to some degree not helpful if it is so readily ignored by the sceptic.

Chemical knowledge has become the workhorse of so much of modern industry through the abilities of its practitioners in providing raw-materials for the transformation of 'stuff' into saleable product. Chemistry assists in the marshalling of similar empirically-discovered transformations into generically valid patterns and of materials with similar properties into groups. Chemistry works within the possible and then forges new procedures to expand on what is possible. It solves physical problems and with that chemistry falls within the 'province of the Natural' (Foster, 1957, p. 18). It would therefore seem unwise to define the 'natural' as being solely that which has been untouched by human interaction. Chemistry is realist and a realistic discipline in that it deals with concrete realities outside of the self (McGrath, 2001, p. 71). Chemistry has been accused of lacking a metaphysical dimension. Would this lack militate against it being capable of informing a natural theology? I will demonstrate below that this too is not the case.

Christianity champions a transformational soteriology, because God in Christ transforms the individual. The wonder that belongs to chemistry is that it is a partaker in and a user of, a 'natural' framework that subtends a transformative ability. Thus the 'natural' I argue for in this book, describes and glories in, a *framework* that enables transformation. Whatever it is that enables that 'framework' is thus the definer of the 'nature' and 'natural' I seek to find the Christian divine imprint in. Such a definition of the 'natural' in a natural theology, immediately sets this present approach apart from certain classical positions, as well as more modern 20th century approaches, for I will be suggesting that the Christ through whom Christians understand they are transformed is potentially on display, or at times could be seen to be visible, within the transformational discipline of chemistry. That which is 'natural' is no longer considered to be just that which is 'of Nature'. Nonetheless if care is not taken, such an approach could be said to represent a partnership between created and Creator as postulated for example in Process Theology, which is not what I am suggesting. I address Process thought further below, yet for now see for example Brooke and Cantor (2000, pp. 315, 338). We must not forget that sin and the Fall engendered a rupture in the framework of creation, such that it 'groans' for regeneration (Romans 8.18–22).

In speaking of a framework, the natural theology I am tentatively moving towards, will tend to move away from a 'God of the gaps' (spoken of by many authors but see W.D. Drees (2002) for a sensitive exposition of this particular term) towards a theology that proclaims the world that the sciences describe, as indeed also being God's world (Re Manning, 2013, p. 124). Such a sentiment chimes with the experience of the Apostle Paul as described in Acts, when on a journey to Athens he sees the Christian God being acknowledged even by those who do not realise that they are worshipping Him (Acts 17.16–34).

Having read thus-far in this section, it might be reasonably asked: 'So which is it? Is God within the natural world or outside of it? If outside, how do you argue God communicates with and within the world? If the Apostle Paul was asking for a re-appraisal of the object of the Athenians' worship, was he asking merely for an intellectual assent?' The work of Vincent Brümmer discussed below provides part of the answer: there is within the orthodox Christian understanding a separation between the transcendent 'wholly other' and the world he created. That being the case, how is God made visible in the 'natural'? These are some of the issues we will expand upon below in the section on natural theology and the Hebrew/Christian scriptures.

Thus to answer my own implied question about the quality of the 'nature' and the 'natural' in the natural theology that I am moving towards, this quality:

- is capable of being perceived by all, Christian and non-Christian (and equally capable of being ignored by both);
- has to do with the frameworks described by the physical sciences, including the mechanism of interactions between reacting components in creating novel outputs;
- moves the focus of our considerations away from individual instances of conceived complexity in nature (established forms of the so-called teleological

arguments or arguments from design, for the existence of God) and towards this rationality just mentioned;
- engenders in those studying it, aesthetic appreciations such as wonder, awe and beauty.

2.1.4 Resolving the Incommensurabilities between Theology and Chemistry

A not unreasonable objection to natural theology might be, as Eric Oberheim and Paul Hoyningen-Huene suggest, that the fields of study comprising 'the things of God' (theology) and 'the things of the natural world' (the natural) are rationally incommensurable: that is that these potentially 'competing paradigms fail[ing] to make complete contact with each other's views, so that they are always talking at least slightly at cross-purposes' (Oberheim and Hoyningen-Huene, 2013, p. 1). In rejecting a Wittgensteinan fideist epistemological approach, I from the theological side, negated this objection. The remark by Oberheim and Hoyningen-Huene relates specifically to the physical sciences and yet it does for this present purpose count as a very reasonable cautioning statement when attempting to formulate or identify a natural theology. The difficulty might be that the chosen approach might function for chemistry for example, but be thought incompatible with natural theological approaches in other physical sciences. In this regard for example Max Jammer (2009, p. 61) describes incommensurabilities in concepts of mass in physics. Far from resolving such conflicts he maintains that such controversies amount to differing views of the development of physical science; no longer are we dealing with right and wrong yet rather we become content to live and work alongside, different models of, in this instance, the concept of mass. We are discussing the same thing, but our point of view and our particular context, lends that 'thing' a different quality dependent upon the context in which it is discussed and utilised.

Furthermore, the way in which Pannenberg's (2008, p. 65) theology combines with his quoted understandings of further work by Jammer is of especial significance in this present study:

> God's immensity and eternity are prior to the finite reality of the world of creation that is the object of geometrical construction and of physical measurement. The infinite space of God's immensity, however, and the infinite whole of simultaneous presence that is God's eternity are implicated and presupposed in our human conceptions and in our measurements of space and time. Thus, God's eternity is different from the time of his creatures but constitutive of it, and his immensity is constitutive of the space of his creatures. (Pannenberg, 2008, p. 69)

which remarks might reasonably constitute an understanding of what a natural theology is, in that God's immensity and presence are implicated in human conceptions (a description reminiscent of Romans 1.19–20). Pannenberg had in preceding pages developed an understanding of 'fields' (analogous to but not synonymous with for example magnetic or gravitational fields in physics) as spacial realities which might be projections of the reality of the Holy Spirit in Creation. Bearing this explanation in mind he goes on to say:

> At this point, I return to the field concept and to the significance of its application to the doctrine of God as spirit. I said before that space and time, or rather space-time, are the only basic requirements of the field concept in the general theory of relativity. Here, the universe is described as a single field, while, in principle, the states of bodily matter (or particles) are considered as singularities of the cosmic field. If all geometrical descriptions of time and space, however, are dependent on the prior conception of space and time as an infinite and undivided whole, the immensity and eternity of God, then this infinite and undivided whole may also be described as infinite field, the field of God's spirit that constitutes and penetrates all finite fields that are investigated and described by physicists, even the space-time of general relativity. This relationship makes intelligible how the divine Spirit works in creation through the created reality of natural fields and forces. The interpretation of the concept of God as spirit in terms of the field concept, then, functions as a key to obtaining some understanding of God's fundamental relationship to the world of nature. (Pannenberg, 2008, pp. 69–70)

He then helpfully, in the same place, adds:

> Such a theological use of the field concept does not and need not rely on any specific field theory that physicists have produced. Nevertheless, it is related to the field language of physics because it claims to deal with the preconditions of any physical field that occurs in the spatial and temporal setting of the universe.

To underline then what has just been said, such a field concept claims to deal with 'preconditions of any physical field that occurs in the spatial and temporal setting of the universe'. The further significance of what Pannenberg has achieved at this point ought also be underlined: he has been able to hold together within the same academic *milieu* or discourse concepts intelligible to theology as well as to those working within the physical sciences: potential incommensurabilities have been dispelled. Furthermore it has been shown that the laws or frameworks which produce the observed behaviours within the discipline of chemistry might rationally be considered as displaying the Christian God's handiwork within the physical world.

2.1.5 The Place of Transcendence in the Natural Theology

We are in this section seeking to understand how the Divine might be visible in the everyday or natural world. The areas of difficulty that must be avoided include holding to the anti-Christian position that God and nature are in fact one and the same on the one hand, against one interpretation of the orthodox Christian position on the other: that humans are estranged from their divine Creator to the extent that they are incapable of discerning God visible in creation. In the first position we see a non-Christian god embodied within the natural world and in the second a radical dualism making any conversation between the two, impossible. Plainly some form of transcendence is needed:

> Against any idea that the natural order was chaotic, irrational or inherently evil (three concepts which were often regarded as interlocking), the early Christian tradition affirmed that the natural order possessed a goodness, rationality and orderedness which derived directly from its creation by God. A radical dualism between God and creation was thus eliminated, in favour of the view that the truth, goodness and beauty of God (to use the Platonic triad which so influenced many writers of the period) could be discerned within the natural order, in consequence of that order having been established by God. (McGrath, 2001, p. 163)

As implied in the introduction to this section 2.1, such a consideration of some form of transcendence matters in the selection of a natural theology because natural theology is vital, as the quotation given above proclaims, for any engagement of theology with the sciences. Furthermore as Christians we contend, we know, that we are speaking about the truth. Many engaged in scientific research would similarly contend that discovering new truths is what urges them on. The Incarnation was fundamentally about the God of unimaginable and indeed limitless power, limiting Himself in His coming amongst us (Phil 2.5–8). Should we be seeking to enable a conversation between the Creator and created, be seeking to discern the Divine in the everyday 'natural' world, we need not look further than the person of Jesus Christ, since as He Himself remarked 'no one comes to the Father except through Me' (John 14.6). Thus both from the point of view of understanding a Christian soteriology (where the non-Christian comes to faith in Christ solely through an encounter with His person as a result of revelation and thus outside the scope of this present project), as well as wanting to have a point of focus for the non-Christian enquirer into the faith, the person of Christ is where that focus should lie. It is in Christ that the transcendence is visible and yet only to the one enabled through Grace to 'see' it.

And how does this seeing 'work'? When Christ asked His disciples 'who do you say that I am' (Matthew 16.13–20) he was met by various answers. The correct answer was not deduced by the one giving it, it had 'been revealed [to him] by my father in heaven' (Matthew 16.17). The powers of reason only led so-far: a radical break with reason was required for the truth to be imparted. This is indeed intelligible when we appreciate that for Christians a transcendent knowledge of God is not a logical extension of either empirical knowledge or of other understandings: the transcendent is not in anyway a continuation of the immanent (Brümmer, 1992, p. 29); there is no logical argument that can rationally derive knowledge of God (although as we have started to acknowledge there are things about God draped around and enfolding the natural world, which leave more than sufficient clues). However, simply because there is no rational or logical derivation through reason alone of the connection between what is empirically observed (the immanent) and what gave rise to it (the transcendent), this does not in any sense mean that there is no connection. Indeed the connection is very real and may simply be *concluded* (Brümmer, 1992, p. 30) through some form of as yet undisclosed insight – an insight we suppose of a similar nature to that which the Apostle Peter received when he correctly appreciated the Christ as reported by the writer of Matthew's Gospel.

Someone might reasonably ask, 'why does God not provide sufficient evidence' for these claims, rather than requiring a plainly non-scientific leap to a conclusion, having observed the evidence of God's handiwork in the natural world. It is in the nature of scientific evidence and indeed of the scientific method itself, that claims should be falsifiable. It is the possibility of a scientific finding being wrong, which drives research and ever deeper knowledge of a given scientific discipline. Yet relationships cannot be founded or rest upon, the possibility of deception. Yes it is true, we are often deceived in our relationships, yet it is our desire that they should be steadfast. In the Christian understanding God as revealed in Christ calls us to friendship with him (John 15.15–17) in 'real-time' and as such *certainty* is required (Hebrews 11.6) for such a properly functioning relationship (plainly we cannot have a relationship with someone we do not believe exists). For ourselves too we require a level of trust in our own relationships for example in the family or even in business: it would be both undesirable and impracticable to be forever questioning whether we even had the basis for a relationship, before starting to work with and in such relationships. At this point then the function of our natural theology is to provide the environment for God to introduce himself to the non-Christian' honest enquirer' so as to commence such a (lasting) relationship, a relationship that is formed according to the pattern of Christ as shown forth in the Scriptures (Brümmer, 1992, pp. 39–40).

Should we therefore examine our Christian (New Testament as well as Hebraic) understandings of the relationship between the Divine and the temporal, we may become enabled to perceive the gap between the Divine and the temporal and so again affirm the nature of the Transcendent. This will then provide a key to perceiving the divine in the everyday, even when the person entertaining that sense of wonder, may not acknowledge the God who has produced this understanding. This incidentally then legitimises the theology being promulgated in this book, as a true 'natural theology' i.e. one which requires no (revelational) knowledge derived through a 'faith', since the Unseen is revealed in the temporal, to a degree anonymously, as Paul implies (Acts 17.23).

McGrath agrees with Torrance that:

> ... dualist assumptions are deeply ingrained within the Western theological tradition, and can be argued to reflect the influence of speculative Hellenistic philosophy rather than its Jewish intellectual context. (McGrath, 2008, p. 14)

In contrast to the Hellenistic world-view that all lays at the feet of the rational mind to uncover, in the Hebraic view, God is separate from the world and hidden (Foster, 1957, p. 42). This is a symptom of His holiness and suggestive that there must be some transformation of the human in order to interact with the divine, as already spoken of above.

The dangers of Cartesian dualism to any *schema* involving a natural theology are evident:

> Then came the age of physical science. The break up of the mediaeval system of thought and life resulted in an atomism, which, if it had been more perfectly consistent with itself, would have been fatal alike to knowledge and society. Translated into science it appeared as mechanism in the Baconian and Cartesian physics: translated into politics it appeared as rampant individualism, though combined by Hobbes with Stuart absolutism. Its theory of knowledge was a crude empiricism; its theology unrelieved deism. God was 'throned in magnificent inactivity in a remote corner of the universe,' and a machinery of 'second causes' had practically taken His place. It was even doubted, in the deistic age, whether God's delegation of His power was not so absolute as to make it impossible for Him to 'interfere' with the laws of nature. (Gore, 1891, p. 73)

Yet in the orthodox Christian understanding, the person of Christ as revealed in the Incarnation was then and as the work of the Spirit within the world, remains today, that 'interference' in the laws of nature. It will be readily understood that Pannenberg is saying something similar, as reported in the discussion of his 'field concept' above.

We now digress briefly to introduce some terminology. Foster (1957, p. 18) builds on the work of Mascall and Marcel to create an essential typology: a 'problem' is something that can be solved and so might be said to belong to the 'natural'; a puzzle is something that might be elucidated by a re-combination of current understandings and so is something of a synthesis; a mystery is 'something fundamentally different': it is that which maintains its ability to impress, to remain hidden, to remain epistemically 'other', 'even when understood' (Foster, 1957, p. 19). Mystery exceeds our comprehension. We may see it, we may understand the physics or the chemistry or the mathematics and yet it maintains its inherent inability to be fully comprehended, and indeed even if it is, it remains 'wonderful'. Foster goes on to draw the distinctions clearly between (Christian) mystery and those who espouse what we might today see as a Fideist approach (Foster, 1957, p. 27) as well as that which we might see as magic and superstition. In so doing we maintain a realist approach in the face of the physical world and are not drawn away into fictions and unrealities.

Foster's notion of 'problem' may validly be re-interpreted as that which is a purely scientific investigation: Foster (1957, p. 21) quotes Schlick who 'proclaims the faith that there is no mystery in the world which can resist elimination'. Foster identifies the root of the belief or 'faith' of humans to wrestle with nature and master it, as coming from ancient Greek philosophy (Foster, 1957, p. 42) which in its rationalism 'vanquished' superstition (*ibid.*, p. 44). He is thus implying that to a certain degree scientism or the unswerving belief that science will answer all necessary questions, is a form of religious 'faith'. This is helpful because with this we have identified our contemporary understandings of that which is naturalistic: it is the academic field the natural theological operates in and so builds upon our discussion of 'nature and the natural' above. Natural theology is thus shown to be capable of operating rationally in the 'realistic' physical world whilst insisting that the wonder it evinces speaks of the transcendent. (Interestingly it would appear that all human effort requires the operation of a 'faith' or of a 'faith-in', of some description. As Edward Wilson affirms, even those opposed to theism, do at times enjoin us to make a 'leap of faith' and assent to the eventual triumph of science (Wilson, 1998, p. 265).

With this we now have to separate-out our understandings about the nature of faith. The person with faith or 'a faith' will arrive at a potential problem with certain pre-formed suppositions. This may to some sound like a return to foundationalism, however let the reader be patient. Those who affirm a scientific world view – Gore's 'crude empiricism' as evinced by Schlick quoted above – have their faith resting on the premise that all problems will eventually be elucidated using the power of the rational human mind, through the scientific method. For such persons there is no dualism since there is no 'wholly other' and all of life and history is a 'problem' which will be solved. Yet as we have seen and shall see again below, mystery persists even when (it is thought to be) understood. And the 'attitude correlative of mystery is wonder' (Foster, 1957, p. 34). This sense of wonder is present even amongst those who have no Christian faith, but only a 'scientific' or more properly a naturalistic faith. Our sense of wonder becomes the suspicion that there is a transcendent reality beyond that which we can perceive. It is this suspicion which through an act of God can become the bridge to the transcendent 'Other': for a human to experience wonder becomes an instance where they begin to perceive the specifically Christian 'Wholly Other'. In order to become a Christian, the terminus of this sense of wonder must be revealed to be Christ Himself through an act of God's self-revelation, yet for our purposes the perception of wonder in our enquirer is sufficient for now. In this *schema* our natural theology is accessible to the non-Christian, not requiring McGrath's dogmatic presupposition. No longer should a Christian interpretation *precede* any engagement with the world in such a public theology:

> natural theology offers a comprehensive means by which theology may address the world, and engage in productive dialogue concerning the legitimation and consequences of belief systems. In a free market of ideas, in which competing conceptions of 'nature' clamour for attention, the question of how the natural order is to be interpreted is of critical importance. Presuppositionless exegesis of the book of nature being an impossibility — as is also the case with the book of Scripture – there is room for a proper and informed debate over how the natural order is to be construed. (McGrath, 2001, p. 303)

Such an approach appears to place too great a burden of divine knowledge upon Christians, leaving theologians interpreting nature rather than God; thus theology addresses the world but the world is not permitted to be addressed through itself by the One who in love created it. Christians must forever be open to the possibility that the *how* – never the why – of Creation will be re-interpreted by the One who made it, the Creator. It is precisely this that makes our enquiry into the wonder that is Christ so completely overwhelming and absorbing, and our conversation with non-Christians a mutual journey of discovery. The implications of this approach are clear: that the Christian does not expect to be able to provide full and final explanations of the natural world; that the Christian expects to be changed by the world since it is the locus of their formation, the disciplining arm (see for example Hebrews 5.7–8, 12.7–13), of their loving Creator; that contingency and a profound sense of underdeterminedness inform all their considerations of the naturalistic elements of the world; above all that we should join Christ in the mêlée that is life, just as he did.

Thus from these perspectives, a natural theology must evince a sense of wonder in the beholder, correlative of a mystery that persists even when the explanation has been uncovered. This sense of wonder, of experiencing the transcendent, grips both the Christian and the non-Christian in our proposed *conversation*, leading both to walk together in a dialogue of hopeful exploration. 'Hopeful' because the Christian knows where this search in wonder will conclude, but only on a journey where the way is discerned 'through a glass darkly' (1 Corinthians 13.12) and where God in Christ provides the interventions. Of this wonder we shall have more to say below.

2.1.6 Considerations of Simplicity and Complexity in Natural Theology

In the past certain types of natural theology have, as one of the so-called 'Arguments from Design', made use of the key finding of complexity in the natural world. The overall objective of this section is precisely not to suggest that the types of natural theology discussed here, are candidates in this present book. I may personally find encouragement in the phenomena being analysed, especially as they make use of scientific discoveries. I may in addition speak of what I take to be the imprint of God in the world, but for the reasons given, they are not suitable candidates for my present purpose. Thus the purpose of this section is to move the candidate natural theology I am developing, away from those based around notions of complexity as found in the natural world.

Given the sheer power of the notion of simplicity in so many areas of the everyday (Derkse, 1993, p. 202) it might seem almost perverse to comment upon *complexity* in the same place as *simplicity*. And yet we are here seeking to formulate a natural theology and so the question might sensibly arise: should we seek for the Christian God in the complexities of the natural world or in its apparent simplicity? Does our notion of the grand, the majestic, the glory of God most ably find its expression in the extraordinary intricacies of the Universe, or do we perceive the Divine in something clear, something pared-down, something simple to observe and comprehend?

Derkse makes the most useful point that there exists 'an intricate relationship between the assumed role and status of simplicity and the ontological suppositions of one's epistemology' (Derkse, 1993, p. 206). A not unreasonable criticism from those who deny the existence of God might be that in accepting this assertion I am perhaps mapping out a pathway to perceive of beauty in the manner God has embedded a sense of simplicity in the natural world. I might effectively then be begging the question of where I am proposing to identify imprints of the Divine in my natural theology, if I were to then also suggest a degree of simplicity within certain areas of chemistry. Yet Derkse goes further and provides justification for such a theistic interpretation of the appearance of simplicity in the natural world. Although Derkse does not call it such, he goes on to define a natural theology by proceeding to state that 'the sense of simplicity and beauty is a sign of the resonance between human rationality and the rationality of the ordered cosmos'. Note that he is here linking a *human perception resident solely in the mind* with a series of concrete physical phenomena. Can there be any other explanation of the bridge between the noetic makeup of the human (such that such a resonance is perceived and 'perceivable') and the inanimate, the non-living rational structure of the cosmos, other than a metaphysical one? Yet that is not all, he then goes on to say that 'the sense for[/of] simplicity is an almost metaphysical intuition for what is fitting and right. From this perspective an indicator of beauty is to be identified with an indicator of truth' (Derkse, 1993, p. 206). We have already elsewhere asserted that we wish to hold to a realist epistemology and Derkse here agrees that to appreciate simplicity in this way, as many scientists do, is to adhere to such a view. Thus I would assert that our natural theology is one that prizes simplicity (and as we shall see, elegance and beauty) not only for its own sake, yet also because of the fact *that it is explicitly perceived as being so by the observer; that the observer has within in them the capacity to react positively to the value of simplicity, of beauty etc.*

Yet what of those arguments for the existence of God, the so-called arguments from design, which posit that certain named structures seen in the natural world are of such intricacy that their very existence indicates that 'there must be a God who created them'? is there a 'degree of complexity' such that an item may have been created 'naturally' and then yet another greater degree of complexity such that the object now must have been created by God? Is it being suggested by those who put forward these arguments that lesser, simpler structures were allowed by God to develop in one manner, but that for others He intervened and introduced them fully formed into the natural world? Or is it that the Divine imprint is more easily seen in the complex structures, yet not so readily in the simple? William Paley in his natural theology was of course addressing this issue head-on. The index to his famous book illustrates this only too well: he writes at length about the arrangement of muscles and the skeleton, of blood vessels in animals more generally. In one place he says:

> these provisions compose altogether an apparatus, a system of parts, a preparation of means, so manifest in their design, so exquisite in their contrivance, so successful in their issue, so precious, and so infinitely beneficial in their use, as, in my opinion, to bear down all doubt that can be raised upon the subject. (Paley, 1881, p. 57)

He is here speaking of the human eye. (He goes on in the same place to explain that it is not merely a case of complexity but also of relationship, in that different parts only make sense within the whole).

The apparent knockdown argument to Paley's work was the mechanism suggested by Darwin which of course proposes random competitive incremental change over lengthy periods of time as the solution to how all this complexity came about. Yet against such an assertion, it proves difficult to characterise randomness. The most that can be said, as William Dembski remarks, is that something is *random to the best of our current understandings* (Dembski, 2002, p. 2). Such a conclusion is of importance in teleologically-derived arguments for the existence of God, and from this for questions about the prevalence of design in the natural world: it is not possible to prove that the emergence of life was a random event.

Yet even if it were possible to prove the opposite, that the complexities of life show evidence of some form of design in the process of their emergence, what has been demonstrated? Perhaps if it could be shown that something was too complex to have emerged according to the known processes of evolutionary change, might this be sufficient to prove that the loving (Christian) God had created that 'something'? Dembski has produced a statistical analysis (Dembski, 2002, pp. 6–10) to show how unlikely it is that living objects expressing coordinated working of multiple co-dependent complex components, could have arisen without the input of a designing 'agent' or intelligent hand. Yet as he concludes, the prevailing culture within the academia of the biological sciences will not allow for the consideration of any other than a Darwinian basis for study, having 'faith' that where current understandings in evolutionary biology are incapable of explaining the emergence of particular complex living systems, a mechanism is sure to emerge eventually to do so (Dembski, 2002, pp. 22, 23).

Yet even if or where, it is possible to demonstrate a mechanism, this is not in itself a comment upon agency. In this way, Alexander Rosenberg indicates that teleological explanations of physical interactions in the natural sciences are non-causal: if we have established or can reasonably hope to establish 'causal laws' Rosenberg believes, any question as to the 'why' of something is 'unfounded' (Rosenberg, 2005, p. 59). He does not however immediately explain what a 'causal law' is: presumably it is a scientific principle with ontological status, something which is itself disputed.

We seem to be going around in circles, evading any clarity of what the perception of 'complexity' might bring to the search for a natural theology.

It seems that demonstrating complexity on its own would not make for a sufficient natural theology to cause the non-theist to examine the claims for the existence of God. According to the criteria set out above, such a natural theology would be insufficiently accessible. Indeed is it truly complexity on its own which brings a sense of wonder? Consider cooled volcanic magma: it is indeed structurally highly complex and appears to be increasingly so with the ever greater violence of the quenching process. If the cooling process is slowed down, crystals have a chance to form. Suddenly there is order to the resulting rock and indeed that order might simply be great regularity as seen in smooth uniform marbles. Derkse speaks of the use of human intuition, the capacity of the human mind to analyse with both synthetic and unifying intent, which tames 'unconnected variety' in the discerning of pattern, of aspects of rhythm and harmony (Derkse, 1993, p. 160). In this way he comments that science has the task of *decomposition* for the purposes of comprehension and *recomposition* in order to make use of it all. We seem to have arrived at a synthesis of simplicity with complexity: human cognition within a purposeful framework enlivened through its unique meta-physical insights, joyfully rationalises complexity as patterned beauty (Derkse, 1997, p. 49), thus establishing pathways to truth. Drawing on our epistemological framework discussed above, perhaps such aesthetic activities as the rationalising of the complex, amount to apprehensions of the Divine: a realistic touching of the beauty of God in the natural world. Simplicity and complexity coexist – a similar instance of incommensurability as discussed above – within the same set of observable entities. Arguments by for example Shook, who accuses theology of having to abandon science in order to explain (ever increasingly observed) complexities observed within nature (Shook, 2010, pp. 98–99), may be set aside in the light of Derkse's work.

2.1.7 The Role of Aesthetics: of Elegance, Beauty and Wonder in Natural Theology

The explanation offered by Darwinism of the origins of the complexities seen in the natural world as mentioned above, cannot explain away the sense of wonder felt at apprehending the intricacies uncovered. The remarkable fact of the human ability to be cognisant of these aesthetic properties and that they are markers for truth, has also been mentioned. John Barrow comments:

> The prospective properties of things [… he includes Beauty, simplicity, truth …] cannot be trammelled up within any logical Theory of Everything. No non-poetic account of reality can be complete. The scope of Theories of Everything is infinite but bounded; they are necessary parts of a full understanding of things but they are far from sufficient to reveal everything about a Universe like ours. (Barrow, 2007, p. 245)

Barrow is thus (re-)locating aesthetic concepts as possibly adjudicators in the field of theories (as of course Derkse also does), and specifically theories of the origin of the Universe. McGrath makes the link even more explicit:

> A strong doctrine of creation (such as that associated with Christianity) leads to the expectation of a fundamental convergence of truth and beauty in the investigation and explanation of the world, precisely on account of the grounding of that world in the nature of God. (McGrath, 2001, p. 240)

McGrath helpfully gives examples in the same place of where researchers have made decisions on the basis of aesthetics which have led to revelations of essential (scientific) truth (again, Derkse comments upon this at length). This is important because an oft-referred-to work by James McAllister (1989, p. 47) places beauty alongside or below rationality in the quest for truth. McAllister (1989, p. 25) appears mistaken in thinking that the discernment of elements of beauty in any selection of competing 'truthful' theories, amounts to irrationality. McGrath demonstrates that this is not the case. McAllister formulated his paper apparently in the belief that aesthetic principles properly and successfully applied could amount to an attack on the rationalist model of science.

In offering an explanation for the potential of aesthetic properties to be successful in being deciders for truthful theories or propositions, Rosenberg (2005, p. 136) offers some supporting reflections from contemporary Bayesian theorising. In postulating a role for the experienced Bayesian researcher in making *a priori* choices that increase the calculated probability of a given proposition, is he not in fact suggesting that the experience of the more mature (in amount of work though not necessarily in age) researcher, is playing a role? And of course such researchers would then be suggesting that it was the aesthetic qualities of their examples, accumulated over time, which then led them to make more accurate predictions about the nature of the solution to the given current problem. Yet the mechanisms for creating Bayesian predictions are shown to be in some sense subjective (and possibly dare it even be said, subject to considerations of aesthetics: Rosenberg, 2005, p. 141).

One of the most modern of the physical sciences, neuroscience, might be thought to be the discipline that could complete Descartes' work of destroying any notion that aesthetics and especially wonder, might lead us to truth. At the end of his neuroscientific study, Kelly Bulkeley in his 'book about wonder' (2005, p. 3), shows that the opposite appears to be the case:

> Whatever feelings of wonder may be lost in the process of analysis, explanation and knowledge formation, the potential always exists for a renewed and expanded capacity for surprise amazement and curiosity. (Bulkeley, 2005, p. 197)

Thus wonder, beauty and elegance emerge again I would suggest, largely unscathed from criticism to be worthy arbiters of truth, when applied through the lens of experience. The adopted natural theology will therefore be making use of such thinking.

2.1.8 The Place of Scripture in Natural Theology

Texts from the Christian scriptures, the 'Bible', have been used above. Whilst recognising that not all authors accept that the Bible is sustaining of natural theology, it is necessary to provide a scriptural basis for the current project. This will be done partly here and partly within the section on theologies in Chemistry below.

This division is necessary because we are here concerned with natural theology within our epistemological framework – a framework that specifically is seeking to put forward a rational case for the Christian faith outside of any special knowledge or revealed truth about the faith.

Yet if it is the Christian faith that I wish to argue for here, then it must plainly be Christian and not simply theist. The question then becomes 'at which point in the discussion – between the Christian and the "honest enquirer" – does "god-talk" become "Christ talk"'? If, using an understanding developed from the work of Brümmer (1992), we are waiting for our enquirer to conclude that there is a transcendent God even if s/he cannot deduce it, and if we are saying that such a conclusion can only come about because God has formed it within them (because 'faith is the gift of God' Ephesians 2.8), it follows that the God our enquirer is responding to, is the God who is (self-)revealed in Christ. Such a response is taking place immediately before God 'names himself' to them. Francis Collins offers a touching and powerful account of this process (Collins, 2009, pp. 3–9): the enquirer is beginning to suspect that there is 'something more' than what they can logically deduce from their natural surroundings, somewhat more than their scientific research tells them is there.

Here we can combine the work of both Brümmer and Plantinga and state that against the positivists, knowledge can and does extend beyond that which we can perceive through our physical senses.

The God being argued for in this natural theology is revealed to us in the Incarnation as the Christ, and communicated to us in and through the Spirit, who in the Christian orthodox understanding was the 'master builder' (Proverbs 8.30), God Himself, at work in Creation at the beginning, in the Genesis 1 and 2 accounts. God is delighting in His handiwork [see also Alan Richardson (1953, p. 50)]. Furthermore as Patrick Sherry remarks, in considering Proverbs 3.16–18 it is noticeable that this pericope in describing Wisdom says:

Long life is in her right hand; in her left hand are riches and honour. Her ways are very pleasant [from the Hebrew נעם : no'am, see below] and all her paths are peaceful. (Sherry, 2002, p. 57)

Although the beauty of God both in classic and contemporary writing is well attested to, the phrase 'God is beauty' or 'God is beautiful' does not appear in scripture, whereas for example 'God is love', plainly does (Sherry, 2002, p. 56). Does this pose a problem? I believe it does not and illustrates how God is written about on account of how He acts towards individuals: God is not often portrayed for example as 'lovely' although God is love. Similarly there is a 'something' about how God is manifest – how God communicates Himself – to individuals under certain circumstances that elicits the property 'beautiful' from them when asked to describe how God should be understood and moreover how His Creation should be framed. This 'communication' is the key: where beauty is spoken of in relation to God in the Hebrew bible it is frequently as He is experienced when those who speak of Him, feel that they are in His presence (see for example Psalm 27.4 and also Berlin and Brettler, 2004, p. 1311). By way of confirmation see also Psalm 73.25 for example, where the one in God's presence speaks of his 'delight' (*chaphets*) at being there, having gone into the temple for the express purpose of communing with God and having previously been unable to make sense of their own predicament (Psalm 73.16–18). Taken together and as a result, the words used to describe being in God's presence are given in response to this underlying cause and those words are multiple: Sherry gives several examples including (using his transliterations) *hah-dahr* (see Psalm 145.5); *tiphahrah* (see Psalm 96.6); *yophee* (see Psalm 50.2) as well as *no'am* as already explained. Thus theologically the witness of scripture is that the beauty of God is a something that is communicated as a result of being in God's presence, in the sense of being blessed by God *with* His presence. As a result, as Claus Westermann explains, beauty in Scripture is perceived as 'event' rather than as 'being' and with that is distinct from for example the appreciation of *objets d'art* (Westermann, 1997, p. 585). And this blessing 'means abundance, wealth, thriving, success, exuberance; and this includes beauty in a variety of ways' (Westermann, 1997, p. 586). It is in the poetic language of scripture, which includes the Psalms as mentioned in Sherry's work above and especially 'in the collections of Proverbs – yet not only there – [that] the description of beauty is expressed by … rich and abundant language' (Westermann, 1997, p. 597). Thus multiple Hebrew words are used to describe beauty, since this sense of blessing, of being blessed by God, is communicated by Him in many ways as illustrated below:

transliteration	Hebrew text	Part of speech	Partial definition	Example of use
Chaphets	חָפֵץ	verb	to delight in, ''take pleasure in, desire, be pleased with	Psalm 73.25
Hadar	הֲדַר	Noun Masculine	ornament, splendour, honour	Psalm 145.5
Tiph'arah	תִפְאֶרֶת	Noun Feminine	beauty, splendour, glory	Psalm 96.6
Yophiy	יֳפִי	Noun Masculine	beauty	Psalm 50.2
No'am	נֹעַם	Noun Masculine	kindness, pleasantness, delightfulness, beauty, favour	Proverbs 3.17

And in the poetry of Proverbs, we see a proliferation of similitudes where beauty is repeatedly compared to wisdom (Westermann, 1997, pp. 597, 598). Expanding on such similitudes Westermann goes on to note that we:

> can find parable and comparison in …. the love song [meaning the Song of Songs], the Psalms, the sayings of the prophets. Every explicated similitude is poetry: an event is compared to one that has occurred in another realm, and thus the one is illuminated by the other or helps to explain the other. (Westermann, 1997, p. 598)

And as Westermann observes in the same place, whilst there are many words used to speak about these events of beauty, they are all in fact illustrative of God's story, his narrative (even some narrative can be seen as poetry, see p. 600), his works, amongst his people.

Since in 1 Corinthians 1.24 Christ is described as 'the wisdom of God' and in Ephesians 2.14 as 'our peace', and since it is the Spirit of God who communicates Christ to the world, it is justified to assert that the Christian Trinitarian God imparts beauty to the ways of Creation [Sherry, 2002, pp. 70–72, and see also Tom Schreiner (2013, p. 299) for a brief discussion around the personification of Christ in Proverbs 8.22 and the Christian New Testament]. As Westermann observes, beauty 'is an intrinsic quality of the creation …. because the creation is beautiful in the eyes of God' (Westermann, 1997, p. 587). Thus God is self-revealed through a blessing: this blessing being the gift of perceptions of beauty sensed in creation.

Crucial to the arguments advanced in this present book is Sherry's further observation in the same place that 'the ascription of divine beauty is made as a result of experiences of worldly beauty, which is regarded as reflecting divine beauty' – occurring in Westermann's 'other realm' (see above). This overwhelming and at times sudden assault upon the senses that is beauty, being God Himself who is being apprehended in this way, is expressed thus in the Psalm 27.4:

> I have asked the Lord for one thing – this is what I desire! I want to live in the Lord's house all the days of my life, so I can gaze at the splendour of the Lord and contemplate in his temple.

In connection with this, what is exciting is that the Psalmist is contemplating this aspect of God in Creation, of Christ, by 'faith' in the sense that this aspect of God's beauty is not seen in the visible. Such an apprehension of beauty is in some measure similar to that same aspect of beauty in chemistry. Such beauty is held in the mind's eye and not by sight, since of course the molecules and processes, the novel compounds, are being appreciated and yet are not seen.

Now both the realism of the beautiful and its being perceived, clash headlong at the vision of Christ, as Jozef Wissink relates:

> Would a theological reflection on the person and work of Christ not be leaving something substantial out, if it were accompanied by a complete refusal to consider the glory and beauty of both the person and the work? [Laat een theologische reflectie over persoon en werk van Christus niet iets wezenlijks weg, als er geheel niet wordt nagedacht over de heerlijkheid en schoonheid van deze persoon en dit werk?]. (Wissink, 1993, p. 11)

This is a rhetorical question that demands a loud affirmative reply. Wissink says this in the context of reminding his readers the words of Isaiah as follows:

> He sprouted up like a twig before God, like a root out of parched soil; he had no stately form or majesty that might catch our attention, no special appearance that we should want to follow him. (Isaiah 53.2)

Thus we should not be looking to an oxymoronic response to the visual tragedy of the flogging and crucifixion of Jesus on Good Friday when beholding His beauty, but rather accepting of the tensions and straining of our understanding of what beauty actually is, as Wissink acknowledges in the same place. Christ in his person creates this *contrast* between what is anecdotally one understanding of beauty, and that truth of beauty which is resolved in Himself. Likewise Sherry reminds us that Karl Barth in reflecting on this self-same pericope also speaks of Christ in Incarnation as revealing the beauty of God (Sherry, 2002, p. 74).

Note the following more paradigmatic view of scripture and beauty:

> The message of the Scriptures is not only that Yahweh is king over his people but also that his people will see the King in his beauty, that they will revel in his promise, and that knowing him will be all-satisfying. The Edenic and paradisiacal love between a man and a woman is the closest analogy on earth to the delights and pleasures of the love that marks Christ's relationship to the church. (Schreiner, 2013, p. 319)

Schreiner later conflates this 'King in his beauty' with Christ as the 'new David' presiding over a 'new Eden' (Schreiner, 2013, p. 363–364).

Thus the natural theology I develop here, being one that accentuates the beautiful, is commensurate with the Christ revealed through the Spirit, in those scriptures. In this way if something within nature is perceived to be beautiful, it is the Christ who gave it form: 'the beauty of Christ makes manifest His own watermark within creation, since by Him and through Him all things were, are, and continue to be' (Ward, 2003, p. 43). If complexity or indeed simplicity (within the meaning of so-called 'arguments from design' or the teleological arguments) are aspects of this beauty, it is Christ that gives rise to these appreciations. If part of a natural world which appears capable of being engineered to create chemical compounds, the ingenuity of which fascinates the chemist that synthesises them, is it not Christ causing this to be so? And of course as Jesus Himself said of the *paraclete*: the Spirit would take from Him and impart it to His followers (John 16.14–15).

Immediately it is obvious that we are speaking of aspects or properties that are so perceived *from our perspective*: things/events/objects are variously simple, complex, beautiful, cause wonder, are seen as intricate – all from a *human perspective,* some thinking they are so and others pleased to quell their enthusiasms in the name of knowledge. Thus and again, I am speaking of matters in the natural world, that I am calling others to appreciate – in the course of our *conversation* – in a similar way as I do. Yet I belong to one of Wynn's 'communities of trust' where centuries of reflection and living according to consistent (orthodox Christian) principles have, through everyday experience, taught us that Christ is alive and active in some real and realistic sense, within the temporal sphere He has created.

We might think of God as 'undifferentiated purposefulness' (see Wynn, 1999, p. 155). For example when we say, after scripture, that 'God is love', He is always thus (meaning He was love in the 18th century and is love in the 21st and in all places). Since I uphold the notion that God upholds, projects, the Universe, it is inconceivable that the truths we know of God (for example that 'God is love'), should not be true and discernible in this temporal sphere, as they are in the sphere that He inhabits, meaning that these must be true at all times. And similarly it would surely be absurd if any vestigial appreciations of the aesthetic welling-up within humans 'created in His image' did not emanate from the Christ who formed them, being in this way communicated by the Spirit who discloses them.

Yet there are difficulties: if 'God is love' some might say, why are there wars or earthquakes? In what way is my natural theology to cope with tragedy? If God is alive and active in this temporal sphere, is He affected by it or not? In what manner is God *like us*, and in what way different; is He 'far away' or is He near? Does our natural theology affect the rocks of planets and the gasses of stars or does it only come into play in the interactions between people? These are important issues because they define the type of Universe we inhabit and inform our expectations of what a non-theistic person, indeed a non-Christian person, might be expected to be able to perceive of God, by simply observing the natural world. The Christian Natural Theologian speaks and the non-theist listens; the non-theist speaks and the Christian listens – one inhabits a space where God is discerned and spoken to (hopefully) daily (or more frequently) and the other at first might think that only what can be tested physically, is what is real and true: the Christian must know and understand in what way their friend is to meet the one they the Christian knows, but which the non-Christian cannot perceive. And so their discourse is *as though there is no revelation,* yet the Christian knows God is speaking *within the conversation.*

And so how are we to understand human contingencies within the Divine certainties of the rules that govern the Universe? Or to put it in another way, how might the eternal never-changing God be revealed and perceived within the dynamic environment which is the chemical reaction or in human relationships?

In his commentary Oliver O'Donovan remarks:

> …. in speaking of the order which God the Creator and Redeemer has established in the universe, we are not speaking merely of our own capacities to impose order upon what we see there. Of course, we can and do impose order upon what we see, for we are free agents and capable of creative interpretation of the world we confront. But our ordering depends upon God's to provide the condition for its freedom. It is free because it has a given order to respond to in attention or disregard, in conformity or disconformity, with obedience or with rebellion. (O'Donovan, 1986, pp. 36–37)

Thus it can be seen that it is our own attitude 'of heart' which then regulates our perceptions of this God-imposed order, indeed of the world around us. We can choose to 'see things as God would have us perceive them' or we may rebel and choose not to. Perhaps we might go so far as to suggest that the non-Christian person, in seeing things as God would have us do, is beginning to be susceptible to revelation (Romans 2.14–16). Furthermore in being ourselves responsible for the ordering *of what we see,* and understanding that this ordering is according to a God-given freedom, it is not *necessarily* Godly. Importantly, the Universe is shown to be 'ordered' only within the perception of the human. And of course O'Donovan has not said here what is also evident, which is that the concept of 'order' within human freedom might itself be an entirely human construct. If this is the case then a person could only have such an appreciation of order, as a result of

God revealing that to them. Thus in the context of the theme of this book, within a single conversation, the non-Christian and the Christian are conversing as if over a wall: the start of a willingness to see the Universe according to a God-given perception leads to a transformation (see Romans 12.2) in our abilities to 'see'. Scripture therefore gives us to understand, indeed leads us to the point of, a new type of 'sight', where all of Creation is seen to be 'of God'. This is the 'appropriate conformity of human response to divine act' (O'Donovan, 1986, p. 36) amongst those of the Christian faith. Thus it should not come as a surprise to hear the Christian arguing for perceptions, indeed interpretations of more generic perceptions, of the natural world, as being a necessary consequence of knowing the God through whom the Universe was created (see Hebrews 2.10).

It is precisely on account of the understandings gained by reading the cosmology of Psalm 8 through this Epistle to the Hebrews, that O'Donovan is able to say that the unknown writer of this quasi-Sermon 'sees in Christ ….. the vindication and perfect manifestation of the created order which was always there but never fully expressed' and importantly for our present discussion that the 'elusiveness of that order in our experience did not mean that it had no kind of existence. It existed from the beginning in God's creative conception' (O'Donovan, 1986, p. 53). As a result we can agree with the Epistle to the Hebrews, that Christ is the first and final cause of the ordered cosmos (Hebrews 2.10 and see also O'Donovan again in same place). The reader will note that I am using 'order' in differing ways here. Order as in 'created order', is plainly used to allude to the Universe or 'Creation' more widely. But O'Donovan's 'order' as used in for example the 'order which God has created in the Universe', has to do with the laws God has embedded in the frameworks of Creation and of which our knowledge is necessarily contingent.

Thus in conclusion on this point we can reliably assert that from a Christian Scriptural perspective, all of the natural world speaks of Christ as communicated to us through the Spirit of God. As Christopher Rowland remarks, scripture testifies 'to the divine being discerned through the ordinary course of nature, whether that be the physical world, or human intercourse and the various modes of human engagement' (Rowland, 2013, p. 28). Rowland then helpfully underlines this aspect of the challenge biblical texts provide to the 'worldy' or normative way of 'seeing': they are 'effective texts … they persuade, disturb, and elicit praise, a sense of awe or injustice … their purpose is to awaken people to life reflecting the divine image, by drawing readers into the dynamic of communication …' (Rowland, 2013, p. 28) – all of which is a renewed encouragement to use the present approach of a natural theology as a means of facilitating a *conversation*. Such a conversation within the adopted natural theological framework once again does not seek to prove the existence of the Christian God, but through (as Rowland remarks) persuasion and challenge encourages our non-Christian chemist researcher to recognise within their experienced sense of awe, something of 'life reflecting the divine image'.

2.1.9 An Interim Conclusion: which Natural Theology?

It would be useful to categorise natural theologies more generally to ascertain where this current study sits in terms of such explorations. This would then allow an appropriate natural theology to be selected as a vehicle for the use of chemistry in an extension or enhancement of that natural theology.

It can be difficult to define precisely what a natural theology is, although as Re Manning (2013, p. 1) has demonstrated this does not stop one going on to edit a large volume on the subject. The objective of this book is to describe the development of a conversation between the scientist and the Christian, more specifically in this case aided by and through chemistry. Such a conversation is not static across the centuries and is re-invented for each generation or change of public philosophical outlook (Casserley, 1955, p. 1). The essentials of the Faith are unchanging yet the manner of their presentation alters for each generation: both for the Christian proposing the natural theology and the non-Christian listening.

From Re Manning we can safely assume that those approaches tackled in his *Handbook* (2013) are indeed natural theologies, at least to some. His categorisations are made thus: Historical, Theological, Philosophical, Scientific and Aesthetic. Having divided all theologies more generally into the revelational and the rational, L Harold deWolf (1958) expands upon Casserley's (1955, pp. 2–3) classification of four broad categories of natural theology (1. an intellectual debate internal to the mind, 2. an argument using observations of the natural world, 3. a theology of nature sufficient in its cogency to convince the non-Christian of the Reality underlying it, 4. relating a similarity between an experience of the natural world and an experience of the world of Christian Faith) by adding a 5th, being the 'philosophical evaluation of doctrines believed to be revealed'. DeWolf mentions William Temple as proposing this particular development, and hence I suggest that it might include for example, demonstrating how social action in showing love of neighbour might rationally be explained as having been derived from the revealed doctrine of Christ sacrificially giving Himself to humanity on the Cross.

If we might briefly consider this classification: an example of type 1 might include Anselm's Ontological Proof; type 2, the Cosmological argument or arguments from complexity; I offer no examples of type 3, although see section discussing some of William Dembski's work below; type 4 would include considerations like those of Pannenberg given above as well as Guy Bennett-Hunter's (2014) treatise on Ineffability and finally type 5, where by way of example I offer the effect upon appreciations of the ways of Creation resulting from a philosophical evaluation of the revealed doctrine of the Trinity. I appeal to this latter form of natural theology as being most likely to make an appeal both emotionally and logically within the context of the proposed *conversation* between the natural theologian and enquiring chemical researcher. This then is how 'natural theology' is understood in the context of this present book. It has the elements as described in the sections above and rests upon the epistemological basis as delineated in chapter 1. It is new in that it does not seek to prove God's existence. It is new in that although being a type of the 'argument from design', it does not concern itself with complexity.

In this way, I would agree with for example Dembski's commentary on CSI (Complex Specified Information in natural systems, most notably biology) that this does show that 'natural causes are incapable of generating [the] CSI [they nonetheless exhibit]' (Dembski, 1999, pp. 153, 170): I find this a convincing argument. Yet my acknowledgement of the worth of such analysis, does not cause me *per se* to enter into a relationship with Christ. I offer this example most certainly *not* because I endorse Dembski's wider thesis and philosophy, but merely by way of illustration that not all arguments from design have an immediate appeal; they can appear overly complex to an honest enquirer and are not as accessible I would suggest as the treatment I am proposing here centred on apprehensions of beauty. Similarly with, as recounted by Richard Southern, Anselm's Ontological argument (Southern, 1990, pp. 127–137) – it makes sense, indeed even Bertrand Russell said this argument 'is to be treated with respect' (Russell, 1961, p. 411), yet it is hard to see how this alone might cause me to fall in love with the Christ who gave Himself for me (Galatians 2.20). Perhaps these are the natural theologies more suited to a quieter age, an age that allowed one to believe that sharply drawn proofs based on the order perceived in the Universe, would generally appeal. It is this sense of the simplicity of an age where such language was used (as in 'sharply drawn'), which 'is passing' (Murphy and McClendon, 1989, p. 191). Nancey Murphy and James McClendon, in a paper which is now seen to be seminal, go on to modify a diagram showing 'modern' theologies in a two dimensional space (*ibid.,* p. 196*)* by adding the third dimension (*ibid.* p. 199), the result being that according to formal vector mathematics, such 'postmodern' theologies may show components reminiscent of variously for example representationalism, collectivism and foundationalism or again of skepticism, individualism and expressivism – for the purposes of this present argument, the detail is less important than the fact that human thought and comprehension of the Divine, is no longer rigid, simple and two-dimensional, but may be multi-layered and complex, even contradictory. Yet overall such theologies show 'holism in epistemology, the relation of meaning to use in philosophy of language, and the discovery of an organic ..... view of community ... – a corporate metaphysics' (Murphy and McClendon, 1989, p. 199) – all of which have been put forward in the chosen epistemology described above.

The beauty of such a position, of such a natural theology, is that it deals with *all* of the person: someone living within a web of relationships, who is possibly both scientist *and* mother, chemist *and* theologian, teacher *and* student, lover *and* beloved, sinner *and* saint, 'both-and' rather than 'either-or' (Casserley, 1955, p. 22): a person Christ Incarnate came to be in relationship with. More than this, it should be noted that Christ was made to be like humans (Hebrews 2.17, 1 Timothy 2.5, see also Brown, Fitzmyer and Murphy, 1990, pp. 1317–1319) and that as such we should be seeking for a certain human quality in our chosen natural theology if it is to be reflective of Christ. Such 'human qualities' Wynn terms 'evaluative responses':

> Contemporary alienation from religion reflects, I suggest, not so much the sense that it lacks evidential support, but rather the belief that it is of no real consequence existentially. And any natural theology which appeals merely to the abstract intellect rather than drawing upon a range of affective and evaluative responses to the world is likely to contribute to this sense that religious belief lacks existential depth.

And for the point of view of this present study helpfully goes on to say:

> Now of the traditional natural theological arguments, it is the design argument which is most naturally associated with an evaluatively engaged response to the world. The ontological argument is after all purely a priori, making no reference to the quality of our experience. (Wynn, 1999, p. 3)

This is precisely the point being made above about the lack of efficacy of certain of the natural theological arguments in formulating a conversation between the Christian and the non-Christian chemist.

As a result I am not wanting to appeal to 'regularity understood abstractly or a quasi-mechanical conception of the world' as I explained in the discussions on order above, but rather to the particular sense of an evaluation of beauty, that 'the world is a locus of value' (Wynn, 1999, p. 15). Of this value, the physicist Andrew Steane suggests in a dialogue reminiscent of those employed by the ancient Greek philosophers:

> [A] We seem to be a long way from the argument from design.

[B] Not so very far. I have just admitted that all arguments from design can only suggest a very restricted amount about God, and then one has the problem of pain.

[A] But you are saying that, if one argues not from particular physical structures, but from the human intimation of meaning and value, then the arguments are at least suggestive.

[B] Yes.

(Steane, 2014, p. 159)

Such a version of the argument should 'not appeal simply to the disengaged intellect, and only as an afterthought' and in addition should have this 'evaluative commitment built into its premises' (Wynn, 1999, p. 15).

And so the version of the design argument – for that is what it is – being proposed here, is not of the traditional type: there is no appeal to intricacy or complexity as I mentioned for example in Dembski's work above. There is by contrast an appeal to the power of similarity (allowing for truth-conducive links to be established between the worlds of science and 'evangelical' or Faith experience, see Casserley, 1955, p. 3) and it references the experience of people in their everyday lives. It is a natural theology that puts forward a suggestive argument (see Steane above) utilising the 'human intimation of meaning and value', specifically in this case the meaning and value attached to intimations of beauty. There is then above all, here posed the question about the nature, the quality of, indeed the wonder of the fact of the existence of, the evaluative response within the human mind. This lies at the core of the natural theology being proposed to be used in a survey of current research in a narrow branch of the chemical sciences.

2.1.10 The Place and Function of this Natural Theology

We have thus-far spoken of the desired components of a natural theology. It has already been noted above that natural theology, as a major evidential anchor for justified belief in God, was previously one of the main methods in use within a foundationalist approach to the justification of a rational belief in God. As such this was the face of justification being presented to the non-Christian public, yet it does not answer the question as to why such an approach might be thought the most efficacious to be presented, from a Christian perspective. In other words, why would the Christian think that a natural theology was the most effective method of engaging a non-Christian in questions about God's existence? Answering this question will now be attempted. Doing so successfully will enable us to link the overall proposed epistemological approach for this project as explored in the first section, with an appropriate Christian theological underpinning.

Consider the familiar opening words of Psalm 19:

> The heavens declare the glory of God;
> the sky displays his handiwork.
> Day after day it speaks out;
> night after night it reveals his greatness.
> There is no actual speech or word,
> nor is its voice literally heard.
> Yet its voice echoes throughout the earth;
> its words carry to the distant horizon.

These words encapsulate so much of what it means for a Christian to contemplate nature and see therein the Divine imprint. Below I develop an argument for linking certain perceptions of one part of the physical world with the God that created it, and yet here in this Psalm written perhaps twenty-five or more centuries ago, the writer has already achieved that objective by linking the declarative 'heavens' with God's law in the two halves of this Psalm. As Konrad Schaefer remarks:

> ... the two halves answer the question, what is the source of revelation? The response is twofold, nature and God's revealed law. The *tôrâh* embraces natural revelation, yet transcends it as it restores the soul and teaches wisdom. The cosmic and moral orders are complementary spheres of God's design; the two can be contemplated in the visible world and within the moral fibre of the heart. (Schaefer, 2001, p. 45)

In reflecting on that Divine imprint seen in nature, it has already been noted above in considering prudential accounts of religious epistemology, that arguments that might be effective in attracting an enquirer to consider the claims of Christianity, will neither be sustaining for that person in the longer term (should they decide to follow the faith in practice), nor be operative in an existing Christian's day-to-day practice of their faith. In considering the function of a natural theology within a person's epistemic understanding, it is thus obvious that there are tensions between what might make such a person ask questions about the Faith, and what provides 'proof' or justified true belief in the God of that Faith.

It is therefore sensible to expand upon what our developed natural theology is, and is for. We have already stated that it is not there to *prove* that God exists. Our rationale for a natural theology has to deal with the growth in faith of one who is already a believer, as well as the complexities of an individual's search for belief. On this schema a natural theology has functions both for Christians as well as 'enquirers' about the Faith. This twin functionality has importance in our consideration of objections to natural theology which we shall address below.

When the father of a boy suffering from what we might perhaps today characterise as a form of epilepsy had a conversation with Jesus about his condition, Mark's Gospel reports the encounter thus:

> Jesus asked his father, "How long has this been happening to him?" And he said, "From childhood. It has often thrown him into fire or water to destroy him. But if you are able to do anything, have compassion on us and help us." Then Jesus said to him, "'If you are able?' All things are possible for the one who believes." Immediately the father of the boy cried out and said, "I believe; help my unbelief!". (Mark 9.21–24)

Plainly as recounted here, the father was not a militant unbelieving person. In the face of a requirement set out by Jesus, to separate the normally accepted course of events (that the boy may eventually and almost inevitably suffer some fatal injury), from the far more hopeful outcome offered by himself, the father struggled to grasp the possibility that the complete physical cure of his son's condition was on offer. This short quotation is offered to illustrate that what is seen, is capable of being interpreted in a variety of ways. McGrath also makes this important point:

> …. apologetics is grounded in the resonance of worldview and observation, with the Christian way of seeing things being affirmed to offer a robust degree of empirical fit with what is actually observed – the "best explanation" of a complex and multifaceted phenomenon. This basic approach can be seen in John Polkinghorne's discussion of the capacity of various world-views to make sense of various aspects of reality, using four criteria of excellence: economy, scope, elegance, and fruitfulness. Polkinghorne here invokes theism as a more powerful explanatory tool than naturalism, and holds that a trinitarian theism is superior to a more generic theism in this respect. (McGrath, 2008, p. 17)

McGrath goes on to show that for some Christian people, a natural theology affirms an existing belief whereas for other believers it might seem to have some 'apologetic potential' (as has already been related above). A natural theology according to this schema does not constitute a 'proof' of God's existence yet rather as the 'best fit' for observed phenomena (McGrath, 2008, pp. 16–18). (We must be careful here not to translate 'best fit' into 'best bet' and alter a belief sincerely held in the truths of the Christian gospel, into something that is effectively one of the prudential accounts discussed earlier).

McGrath however goes beyond the bounds of an intellectual assent to the claims of a Christian natural theology by saying:

> More recently, the waning of modernity has provided a congenial context for the liberation of natural theology, so that its deep intrinsic appeal to the human imagination may be realized. Natural theology is to be understood to include the totality of the human engagement with the natural world, embracing the human quest for truth, beauty, and goodness.

We invoke the so-called "Platonic triad" of truth, beauty, and goodness as a heuristic framework for natural theology. When properly understood, a renewed natural theology represents a distinctively Christian way of beholding, envisaging, and above all appreciating the natural order, capable of sustaining a broader engagement with the fundamental themes of human culture in general. While never losing sight of its moorings within the Christian theological tradition, natural theology can both inform and transform the human search for the transcendent, and provide a framework for understanding and advancing the age-old human quest for the good, the true, and the beautiful. (McGrath, 2008, p. 19)

Now perhaps we might start to comprehend perhaps both what a natural theology is and is for: it is an answer to the wonder felt by those who take the time to contemplate the natural world (see also Wynn, 1999, p. 156 where we see this wonder presented as Christian worship, now shown to be a rational response to a 'designer' who provides a 'causally effective summation of the nature of existence', whilst also 'Comprising a synthesis of the perfections which are manifest in creation'). There is something about the human condition that allows for a person to have an 'intrinsic' response to the glories of nature. [Intriguingly for the present argument, Pierre Laszlo (2003, p. 12) tells us that 'Chemistry …. might be defined from the wondering at change', something we will have cause to reflect upon below. Speaking of the syntax used by this commentator, given that he says elsewhere in the same article: 'About ten thousand protein structures populate already this world' (*sic*) I believe it safe to assume that the first quote might respectfully be emended to read 'Chemistry …. might be defined *as* the wondering at change', which assumption also fits the context.]. And for later use in this book we might also note McGrath's apparently sympathetic acceptance of a distinctly platonic ideal.

Thus in conclusion here, the natural theology I am developing takes the chemist's 'wondering at change', mixes it with the sense of beauty that they feel at what they have discovered and then directs these thoughts towards the notion that [a] God is responsible. Yet I have gone beyond this: it is not only that God is responsible but that the Christ through whom the Universe was made, can ultimately be perceived in those beauties of Creation. Thus the *telos* of the natural theology is the hope that it might become revealed theology, that God might graciously reveal Himself to such an enquirer. Such a journey however cannot start and end with natural theology. The latter as I am proposing it here starts for the chemist with questioning at the wonder and the beauty, and ends with a conviction that whilst they may not agree, at least the natural theological argument as presented is epistemically rational and intelligible.

2.1.11 Christian Objections to Natural Theology

In this book we are seeking to inform a natural theology with insights from the chemical sciences. Yet not all Christian authors agree that natural theology is a suitable epistemic strategy for an exploration of the Christian faith. It has already been noted that those in the Reformed Epistemology school avoid natural theology. Why might they think in this manner and in what way could these and similar objections to natural theology impinge upon this project?

The Enlightenment sought to anchor knowledge in the efforts of the individual to rationalise and to reason. It wanted humankind to be free from – as it saw it, the – ancient hegemonies of religion and patriarchalism. In so doing it forced religion to conform to its methodologies and adopt a schema of rationality in its defence of the right to believe in God. As a result this 'right to believe' became the pressure to prove God's existence, because the Enlightenment taught that only what one could prove, could test, was worthy of consideration. Thus a natural theology became the basis for such a proof of God's existence. I have already indicated that this need to prove God's existence is not a fruitful object of this study because theologically such a proof would not and could not by itself be sufficient to lead to faith in Christ. This form or type of natural theology specifically is therefore not useful in this present book. From a Christian perspective, such an argument, even if it could be produced, would not result in 'faith towards God' (Hebrews 6.1), since salvation is God's gift by grace (Ephesians 2.8) and is not attained as a result of a reasoned encounter utilising human arguments (see also 1 Corinthians 1.18–25 which would further appear to rule out salvation through reason alone).

And yet this project affirms the desirability of support for many of the varied forms of natural theology as vehicles for alerting enquirers to God's (possible) presence and actions in the world.

Putting the arguments for a natural theology in this way – that it points people towards God – is an attempt to circumvent certain Reformed objections to natural theology. Consequently this section of the project is attempting to accurately locate a viable natural theology within our epistemic framework. As such we are only interested in reviewing these Reformed objections in so far as they permit us to attain that goal of identifying a natural theology.

Michael Sudduth has provided a helpful survey of these objections. As he says 'reformed criticisms of natural theology have typically not targeted the project of natural theology as such but rather a certain construal of this project' (Sudduth, 2009, p. 38). He goes on to tell us that Reformed thinkers (rightly in my view) reject any enterprise, as I have just remarked, which seeks to enable a knowledge of God that is achieved or obtained outside of the grace of God in Christ: they maintain that coming to faith in Christ is enabled at God's initiative. Thus the crucial point here is that 'grace and truth came about through Jesus Christ. No one has ever seen God. The only one, himself God, who is in closest fellowship with the Father, has made God known' (John 1.17–18). Sudduth does not here place objections to the use of natural theology, voiced famously by Karl Barth (1946), and see also McGrath, 2001, pp. 267–272, in the context of the German Confessing Church of the 1930s where Barth was attempting to hold onto orthodoxy in the face of a National Socialist attempt to construct a religion of their own that might 'deceive even if possible the elect' (Mark 13.22). Colin Gunton delineates Barth's objections rather elegantly:

> Barth is rejecting the beliefs of those proponents of natural theology whose procedure presupposes that there is between divine and non-divine reality such community of nature that the knowledge of the former can be read off the latter without particular divine initiative. (Gunton, 1978, pp. 153, 154)

And thus the argument here must employ a degree of subtlety: yes, Barth was emphatic in his rejection of natural theology but we need to accurately understand the *milieu* in which the discussions were taking place and the type of natural theology under consideration. In this manner, Barth conflated an acceptance of natural theology as being that which could be 'decisive' in someone's decision to become a Christian, with Emil Brunner's apparent acquiescence to the aims of various groups of National Socialists who were calling themselves 'German Christians' in the Germany of the early 1930s. As is entirely consistent with Barth's own soteriology more generally, salvation is through Christ alone – only a revelation of Christ is decisive in that sense – and not arrived at in a religiously syncretic manner within German history as well. Thus as Christoph Dahling-Sander points out both through an analysis of Barth's and Brunner's published theological views on the matter (Fraenkel, 1946) as well as through a reading of Barth's personal correspondence of the time, these matters 'are thus not capable of being separated from each other' (Dahling-Sander 1999, pp. 12–13 and see also McGrath, 2001, p. 281). This is unfortunate in our present discussions because as Barth himself makes clear 'natural theology is always the answer to a question which is false if it wishes to be "decisive". That is the question concerning the "How?" of theological and ecclesiastical activity' (Barth, 1946, p. 128). As Richard Burnett explains, Barth uses 'theological activity' in the sense of speaking of, interpreting, divine revelation (Burnett, 2013, p. 152). What it would appear rational to draw from this, is that Barth is saying that even the very act of speaking or considering natural theology as a means or mechanism for understanding or gaining knowledge of God is a profound error, *where this is being constructed as a means of entrance into salvation*. And as I have said earlier, I am in complete agreement with this. However to deny *any* involvement from natural theology in the movement of the person from non-Christian 'enquirer' to an individual to whom Christ has revealed Himself, leaves the Christian conversationalist without an answer when speaking about the facts of the natural world. 1 Peter 3.15 calls us to have just such an answer.

Thus I contend that a natural theology should neither be born out of a fear of, or to do battle with, the surrounding culture yet rather as a reasonable response to the Divinely imprinted realities of the world around us, suitably 're-contextualized' as Sudduth implies in the same place. I say this since Sudduth appears to follow McGrath in suggesting that a purpose of natural theology is to act as a 'rational preamble to dogmatics'. Constructed in this way a natural theology is primarily a tool of the Christian faithful: useful as a 'dogmatically situated activity of rational reflection on the Christian God' as well as a tool for the 'apologetic deployment of theistic arguments' (Sudduth, 2009, p. 62). Developing and extending Thomas Woolford's helpful analysis of Renaissance natural theological approaches, Sudduth and McGrath's theologies might be characterised as *pessimistic and post-fideal* (meaning after having come to faith as a Christian) whereas in contrast I am advocating a more *optimistic and pre-fideal* use of natural theology (Woolford, 2011, p. 197). Consequently a solely *post-fideal* use of natural theology would not satisfy the honest non-Christian enquirer (since the apologetic arguments are being deployed for combative reasons and specifically not to invite anyone to follow Christ) and so do not have a great appeal in the current project. Furthermore it is unclear how such a natural theology could be 'recontextualized' (see above and assuming this term to mean 'an enculturalised re-working to make it intelligible and acceptable to a contemporary non-Christian audience') whilst at the same time be held for a Christian audience alone. Thus whilst I applaud the epistemic tools provided by Reformed reflection on Christian faith, it would be sad indeed if a natural theology could *only* be employed by the faithful either for personal edification or for refutation. Instead I suggest that a natural theology can be used to encourage the suspicion that the ability to appreciate the aesthetics of the created world, has arisen from beyond the physical realm, again in accordance with 1 Peter 3.15 as suggested above.

A further component of the Reformed objection to natural theology has to do with the understanding that sin has so corrupted the vision and perception of the non-believer that such are utterly incapable of any appreciation of the Divine. This latter argument is based on a Reformed interpretation of the Christian scriptures and it is therefore to these that we must turn in order to refute it. In the first instance scripture allows for the possession of a conscience by those who are outside the faith. This directs them to act in a way commensurate with God's laws:

> For whenever the Gentiles, who do not have the law, do by nature the things required by the law, these who do not have the law are a law to themselves. They show that the work of the law is written in their hearts, as their conscience bears witness and their conflicting thoughts accuse or else defend them, on the day when God will judge the secrets of human hearts, according to my gospel through Christ Jesus. (Romans 2.14–16)

Earlier in Romans Paul affirms that 'all have sinned and fallen short of the glory of God' (Romans 3.23). Therefore we might reasonably ask in the context of a natural theology, can anyone perceive of God in the natural world, given that we are all sinners? And in the parable of the tax collector and the Pharisee we appear to have the answer. Here, for the benefit, as the scripture itself says, of all those who believe themselves to be righteous and in consequence look down on those they perceive not to be, God declares that the tax collector went home 'justified' (Luke 18.14). The question naturally arises 'justified by whom' to which the answer is plainly God himself. What is the reason for this? It is because: 'for everyone who exalts himself will be humbled and he who humbles himself will be exalted' as we read in that same verse. It is an attitude of the individual, the person, which is an enabler of justification before God, precisely not because of any ability on account of the person, yet rather because God counts an appeal to Himself to their credit. This appreciation or understanding of, the place of the attitude of the believer, would appear to be confirmed when scripture says further:

> because if you confess with your mouth that Jesus is Lord and believe in your heart that God raised him from the dead, you will be saved. For with the heart one believes and thus has righteousness and with the mouth one confesses and thus has salvation. (Romans 10.10)

Thus the ability to *objectively* perceive of, to apprehend, God as creator within a situation, has in part at least to do with the (*subjective*) attitude of the enquirer. The effect of sin is not to *completely* destroy the abilities of the non-Christian to perceive of God in the natural world as certain reformers and their followers teach, yet rather as Andrew Davison points out, to severely diminish or darken the human intellect such as to make the ability to perceive God in Christ in this way, less capable (Davison, 2013, pp. xiv–xv). From this, and with the Reformed objections in mind, it can be seen that it would be quite wrong to deny completely the ability of an enquirer to perceive of the divine within some aspects of the physical world.

A somewhat more nuanced Christian objection to natural theology similarly relates to a Christian interpretation of this 'unknown' God, for if we suppose that God is 'unknowable' (see for example Ps 139.17: 'How difficult it is for me to fathom your thoughts about me, O God! How vast is their sum total!') then surely, as Douglas Hedley relates, a natural theology becomes impossible (Hedley, 2013, p. 582). Yet we are not left in any doubt on this point. In the final part to the prologue to John's Gospel (John 1.14–18) we see this problem both stated and resolved: indeed no-one has even seen God and He is in this sense unknowable, excepting that in Jesus Christ, a man in history, God is indeed made known. This objection to a natural theology is in fact resolved through the proposal that it can only be viable when seen through the person and work of Jesus, 'who is God' (see John 1.18).

2.1.12 Atheist Objections to Natural Theology

We have already outlined above something of the framework of objections to Christian belief itself, demonstrating that these are not always based upon an objective search for the truth. Here we seek to look at such objections as they relate specifically to natural theology.

Let us consider an example. A Christian and an atheist are each involved. Let's assume for now that they are equally qualified and experienced in the fields under investigation.

Both are members of a team that manages to synthesise a complex organic molecule possessing a large number (> 30) of chiral (fixed orientation in dimensional space at junction points) centres. The actions of this novel chemical compound on certain live animal test systems are described. The latter involves the compound binding preferentially such that the progress of a certain disease is dramatically slowed. The process whereby this happens is described in great detail.

The atheist will be satisfied that the explanation of the success of the process provides (accepting the contingencies inherent within the scientific method) a sufficient, cogent and rational account of the actions of the new compound and is satisfied that in demonstrating such a pathway, no theistic explanation is wanting or necessary. For an example of such a common generic 'naturalistic' approach see Matthew Bagger (1999, p. 13), where interestingly any explanation beyond the physical must be untrue, because it is to him 'unimaginable'. From this it will be clear that Bagger is – probably – unintentionally invoking Plantingan warrant to justify his scepticism. The Christian, possessing the same understandings, becomes overwhelmed with the coherence of the process. No amount of explanation can rob him or her of the sense that there must exist *a further account* of the reasons for the natural world to give rise to this level of success. The atheist will counter by providing a step-by-step inferred narrative of the development over history, of the framework for the process: the cells, the compounds, the overall chemical environment within the animals. The atheist believes such an account provides a sufficient explanation of the physical process: no further explanation is necessary to understand the origins of the framework. The framework itself was an inevitable result of the processes over many many millennia following the Big Bang which is itself inferred from other investigations within physics. Thus and again, no natural theology is needed since no explanation is required other than that provided through experimentation. Furthermore, no matter the level of complexity or degree of ingenuity observed in the natural world, an apologetic method which contends that all questions can be solved within *the mechanism of explanation* that has at its core the immense time-periods postulated currently for the age of the known Universe (currently thought to be in excess of 13 billion years), implies an attitude of scepticism which is unlikely to be dissolved by ever greater levels of observed intricacy. This *mechanism of explanation* here has as its *locus*, the atheists' required leap of faith as already quoted above (Wilson, 1998, p. 265). (We remember the work of Dembski quoted above which tells us that the observed levels of complexity in certain living systems where several sub-systems, each of which is itself highly complex, combine together to provide the overall observed functionality, preclude a naturalistic explanation for their development). Christians, within the proposed natural theological *conversation,* and in combining aspects of both revealed religion – their experience of Christ in and through them in the everyday – as well as a nuanced natural theology being the joy of God's *sophia* in Creation *backscattered* in the natural world of today, affirm within themselves 'Christ the [living] hope of Glory' (Colossians 1.27). The atheist and the Christian are respectively placing their faith in different places and objects.

2.2 The Place and Relevance of Theologies of Nature

It is probably going too far to suggest that the two terms 'natural theologies' and 'theologies of nature' are two entirely different disciplines that happen by accident in the English language to share names with a similar arrangement of letters of the alphabet, but at least that will enable the reader to acclimatise themselves to the notion that they really are two very different areas of study.

Natural theology starts with the world as we might see it and aims to point us towards God. A theology of nature, as Ian Barbour explains, is a work of the Christian who seeks to take a known theological stand point and then interprets what we see of nature in this light and often then attempts to formulate proposed courses of action, resulting from these insights (Barbour, 1990, p. 26). There is not necessarily any point of contact between the two. Thus theologies of nature might for example ask such questions as 'what should the Christian response be to the currently ecological crisis?', or again 'how might a feminist theology re-shape our attitude towards nature?'. Both disciplines might be accused of attempting to re-shape orthodox Christian perspectives and understandings. Certain forms of natural theology might re-shape our understanding of God. Process theologies could be an example of these. A theology of nature which recognises the seriousness of current ecological concerns might for example wish to re-interpret traditional understandings of the apparent instructions in Genesis to 'fill the earth and subdue it' (Genesis 1.28). A theology of nature assumes that God is there, whereas a natural theology would start at least from the standpoint that the existence of God is what we are seeking to elucidate.

Having said that there is not necessarily any connection between the two, questions of theologies of nature might well emerge from the sort of conversation that I am proposing in this book. Thus for example should our honest enquirer be at least interested in investigating the claims of the Christian faith further, their thoughts may turn to those Christians who have historically been responsible for the overuse and exploitation even, of the world's natural resources. Such exploitative actions have frequently led to the oppression of local peoples and the permanent destruction of habitats and a consequent loss of species. The natural theologian might then be challenged to explain how those who follow this religion which claims to speak for and with a God of beauty as revealed in chemistry, could possibly be quite so greedy and rapacious. The very answer might well be to show our enquirer that there are indeed theologians formulating theologies of nature which enjoin Christians to respect that which God in His bounty has given us in nature; they might add that 'filling the earth and subduing it' has to do with a God-directed stewardship of the earth rather than the nihilistic destruction of it. These latter thoughts would then be wrapped-up into, or packaged as a particular 'theology of nature'.

Yet it is right and fitting to observe that the loci or focus of such a conversation has only arisen out of the desire to engage with God as well as those who apparently believe in Him and therefore in some sense represent Him on earth. Thus a theology of nature becomes a public enterprise once the relevance of the Christian Church is acknowledged. If the Church and the Christ it proclaims is seen as irrelevant, a theology of nature remains a private affair within the Christian community. From this it can be seen that politically, theologies of nature might be understood as part of the need to be seen as 'relevant' in modern societies. In this manner we are urged to be ecologically friendly and socially responsible in all aspects of communal life.

Furthermore it might be argued that the existence and eagerness by local Christians to implement theologies of nature might itself be a useful apologetic tool in the sense that an outsider to the Church might legitimately conclude that they want to be part of a group that champions concern for the environment, for example, or to join a group of people who demonstrate their passion for a particular commitment in a practical way. This type of influence exerted on others should not be viewed with any form of cynicism since Church groups are often quite genuinely concerned about these issues and moreover their own economic power, whilst not always huge, does nonetheless make a useful contribution to the progress of such ideas. In the way described, a theology of nature has become something of a natural theology.

Similarly a person who becomes a Christian on account of such an influence might choose to continue to have that interest as one part of the outworking of their new-found Christian faith, after they join a particular local Christian church. In such a case a natural theology has had some influence on a particular Christian's own 'theology of nature'.

In neither of these cases, of the intermingling of natural theology and theologies of nature, could these be described as necessarily carefully worked-out and intellectually rigorous treatments of either discipline, but at least they illustrate where the two may become embroiled, one with the other.

In conclusion then, a theology of nature would not be of great use in this present project, yet the natural theologian would do well to be aware of contributions in this area.

2.3 Conclusion

The form of the conversation that is the objective of this book, is one that conforms to the norms of natural theology in that it presupposes no special revelation of God and relies upon that which might be seen and perceived in the physical world. Yet this natural theology is not of the established pattern in that it does not seek to prove the existence of God. Instead the arguments put forward in this book by the natural theologian in their conversation with the chemist researcher are suggestive arguments. They seek to suggest that the human capacity for perceiving intimations of value, specifically in the perception in this case of beauty, has an origin in Christ. This natural theology is drawing our enquirer towards the point in our conversation where they may draw the conclusion – not reached through a logical evaluation – that God in Christ is the source of all that they see, evaluate and empirically test-for, in the natural world.

A natural theology, in representing 'more faithfully than the other traditional proofs the reasoning of the "ordinary believer"' (Wynn, 1999, p. 3) should reasonably be expected to meet the ordinary [dis]-believer precisely at their point-of-need. One of the points of connection between the epistemological position developed in chapter 1 and the natural theology here in chapter 2 is the experience of such ordinary believers in the everyday. In chapter 1, I developed the form of reliabilism being proposed, by including the collective witness of a group of (ordinary) Christian believers or church as part of the process of establishing the correct environment for justified belief to arise. Here in chapter 2 those of this 'church' are being offered a natural theology commensurate with their needs. Chapter 1 shows how for example 'religious experiences' perhaps in the perception of something as being beautiful, may rationally be said to be from God. In chapter 2 we find that the human perception of certain value constructs, including those speaking of beauty, may form part of a natural theology. Our epistemology speaks to our natural theology. For these reasons epistemologically, a natural theological approach offers the most efficacious strategy in our proposed conversation.

The life of Christ intersects a natural theology in the here-and-now, causes to collide the living God in the face of Christ, with the story, the current life, of the enquirer: it is in Marshall's terms both a *pragmatic and a correspondence thesis*. A natural theology consists in seeds of doubt imparted to the mind of the non-Christian enquirer, to the effect that their current naturalistic understandings may not be a valid account of existence after all.

Furthermore, in the context of a conversation, our natural theology provides a rational explanation for the wonder felt by many at the uncovering of the fabric underpinning the natural world. It is a form – loosely perhaps – of the Design argument and allows links to be made between the worlds of science and the disciplines of Christian theology. As I have already said, the natural theology described here is not designed to lead an enquirer to the certainty of faith towards God in Christ through logical reasoning. Such a natural theology instead makes it more likely that a revelation of God towards them will be accepted. It is quite distinct from a theology of nature. This natural theology is overall accessible, life-affirming, outward-looking and engaged with and in the world. It is both optimistic and realistic.

We have established an epistemology for justifying belief in the Christian God in chapter 1 and a natural theology of a type that fits with the epistemology in this present chapter. We now move to an exploration of chemistry in chapter 3 to investigate how certain parts of this very wide discipline, might fit within this natural theology.

Chapter 3: Chemistry and Natural Theology

3.1 Appreciating Chemistry: the Historical Context and Contemporary Understandings

The title of this project derives in part from one of the Gifford lectures 'Reconstructing Nature: the engagement of science and religion' which far from wanting to argue for a particular conclusion sought instead to reinvigorate the interaction between science and religion (Brooke and Cantor, 2000, p. x-xi) by re-examining past science as historians, through the lenses of more recently developed disciplines including psychology, sociology, linguistics etc. The authors reject any 'master-narrative', describing the views of the founder of the Gifford lectures as anachronistic. They are keen to stress how both science and theology change over time, as do the very subjects the founder urged the lecturers to address. They describe the context the Edinburgh lawyer Lord Gifford came from, as being 'highly religious'. In their first chapter the authors attempt to demonstrate that there is no hegemony of one over the other, of science as against theology. By their final chapter they are speculating on the effect on the future of their discipline, that of history and the historian watching the debate, of directed genetic change in the speculator, that is, the human agent. Thus the overall effect is to present an intellectual field that is shifting, as the field of view itself is shifting.

Yet why should any of this matter in the current project which is enquiring whether chemistry might validly inform natural theology?

In their conclusion to their Gifford lectures, Brooke and Cantor, having in their final discourse made some illuminating remarks on the potential role of chemistry and natural theology, observe:

> One reason for offering these chemical snapshots is that they expose some of the difficulties that arose in integrating an interventionist science [chemistry] with a contemplative theology, but also how the difficulties were overcome. In contexts where the meaning and scope of 'nature' became progressively blurred, so the scope of a natural theology would become increasingly problematic. But what we have also seen is that one kind of theology might survive – the kind that sees in the alleged improvement of nature a collaboration between human beings and their Maker. (Brooke and Cantor, 2000, p. 338)

Following certain observations about the dependence of chemistry in the 18th century (and presumably earlier) upon medicine, Brooke and Cantor go onto suggest that chemistry 'became less propitious as a resource for religious reflection' (Brooke and Cantor, 2000, p. 339). In reviewing the chemistry of the nineteenth and on into our own centuries, the authors continue in a somewhat forlorn manner. After hearing from the theologian Gordon Dunstan who suggests that human actions in the evolutionary process are necessarily acting with God, Brooke and Cantor remark:

> Were there still reputable scholars in the late twentieth century [when they themselves were writing] who would interpret that alignment as alignment with the purposes of God? Or was the rhetoric of a secularised natural theology totally and invariably secular? Our future historian would find that theological essays on the subject had not dried up. (Brooke and Cantor, 2000, p. 340)

These are not encouraging sentiments: rather than flourishing, theists are to their mind speculating in a manner that cannot be proven. Taken together, these remarks suggest that chemistry and Christian theology do not easily interact.

As Brooke and Cantor have related, this has to do with chemistry seeming to tamper with nature. Of the practice of dyeing garments Tertullian wrote at some time between 190–220 CE in *De cultu feminarum*, Book 1.8:

> Quis enim est uestium honor iustus de adulterio colorum iniustorum? Non placet Deo quod non ipse produxit; nisi si non potuit purpureas et aerinas oues nasci iubere. Si potuit, ergo iam noluit; quod Deus noluit utique non licet fingi.
>
> [what legitimate honour can garments derive from adulteration with illegitimate colours? That which God has not produced is not pleasing to Him, unless He was unable to order sheep to be born with purple and sky-blue fleeces! If He was able, it follows He was unwilling: what God willed not, of course ought not to be fashioned]. (Tertullian, 2015) with thanks to Prof Dr Claus Jacob for drawing this to my attention.

Any transformation of an object into something else, something it was not previously, something 'unnatural', was to be avoided. And again, what might appear to be attacks on alchemists from for example the 14th and 15th centuries whether found variously in a papal prohibition or literature or again in the graphic arts, can today equally be interpreted as attempts at shielding the general populace from charlatans. Yet as Joachim Schummer says, the overall effect would have been to transmit the impression that these were matters best left well alone (Schummer, 2015).

Yet by the 17th century, of the Irish polymath Robert Boyle (1627–1691), J.J. MacIntosh and Peter Anstey tell us:

> Boyle was one of the leading intellectual figures of the seventeenth century. He was a dedicated experimenter, unwilling to construct abstract theories to which his results had to conform. (MacIntosh and Anstey, 2010, p. 1)

And then later in the same work:

> Boyle's scientific range was wide. Besides his well known work in mechanics, medicine, hydrodynamics and a wide variety of experiments with his vacuum pump, he was interested both theoretically and practically in alchemy …. where his interest seems to have been fuelled more by his constant desire to acquire knowledge of God and the world than by any desire for riches. (MacIntosh and Anstey, 2010, p. 2)

In proof of which we may quote Boyle himself from volume 2 of his papers and folio 57v–58r:

> ….. So that to prove the existence of God from the idea he has impressed on the mind is not to prove it really a Priori; since that idea is not the cause but the effect of the divine existence; though I am willing to grant that in a qualifying sense, this knowledge we have of God by this idea or stamp may be said to be a provision in regard that we obtain it not by the consideration of those effects or productions of God that are without us and made up the visible world. From the contemplation of whose vast extent, regular motions, and admirable contrivance; philosophers and other considering men have in all ages, as from so many manifest effects, inferred the existence of a first and divine cause and consequently have drawn their conclusion by that way or argumentation that all men allow to be framed a posteriori. (Boyle, 1660)

Thus showing that Boyle derived his conclusions after investigation (or 'contemplation') and experimentation. MacIntosh and Anstey also tell us:

> Convinced that Christianity was the religion instituted by God, Boyle was concerned that the Bible should be widely promulgated and he devoted time and energy to having it translated into a variety of languages …. (MacIntosh and Anstey, 2010, p. 3)

Here we have in one of the foremost minds of his age, a keen experimenter using fluids and reactants in a manner that would allow us today to call him a chemist of sorts, being also a fervent Christian. Allan Chapman (Chapman, 2008, p. 20) says of him 'an awareness of the mystical, Divine causality of nature formed part of the very essence of his intellectual being' and that he made 'scientific research an act of worship in its own right' (Chapman, 2008, p. 23) – something from which he perhaps differed but little from contemporary Christian researchers. Thus from an epistemic point of view we have a Christian believer, whom we presume was such from a relatively early age (meaning that he accepted certain Christian truths *a priori*), also able to engage in a certain scientific skepticism and so infer (scientific) propositions *a posteriori*. It is the combination of these two positions which is most interesting, as we have already noted in the work of Foster above.

In something of a pattern that continues to repeat itself in our own times, Chapman (2008, p. 21) says of work in the fifteenth and sixteenth centuries by many investigators: 'chemical thinking had in many ways become more diversified and labyrinthine as a result of all the new classes of phenomena – metallurgical, medical, botanical, and such – that were coming to light, and that were now begging a coherent explanation'. Of the sixteenth century he says that experimental chemistry exerted an 'intellectual fascination' – across Europe – over the men of that age amongst whom he lists in addition to Robert Boyle, the philosopher John Locke and the researcher Robert Hooke.

That 'intellectual fascination' in our own century has altered somewhat. It is no longer the preserve of a few and no longer always the preserve of a single discipline. Trevor Levere suggests towards the end of his survey of the current state of chemistry 'Chemistry and physics ….. have blended into one another in several areas' and 'the way that chemists define their science in relation to other sciences is in the end a piece of territorial assertion' (Levere, 2001, p. 182). He rejects a reductionist approach – one which sees for example a science such as chemistry being reduced to mathematics and physics – either between chemistry and physics or between biology and chemistry and instead says 'chemistry is the only science that now builds or creates much of what it goes on to study'. From the latter it is perhaps easy to see why Process Theology (which we discuss below) is thought by some contemporary authors to be a promising [theological] route into the science. In addition, given the apparent constant threats of reductionism, such a statement also adds detail to our definition of chemistry. In this manner, not only is chemistry about the science of transforming matter, it powerfully has to do with the creation of new materials to the extent that any process that does, may be called a 'chemical process'.

As Michael Weinberg, Paul Needham and Robin Hendry remark, now in the 21st century, chemistry is understood as the 'study of the structure and transformation of matter' (Weisberg, Needham and Hendry, 2011, p. 1), and specifically as the 'transformation of matter from one form to another', whilst continuing to recognise that the discipline can also consist of 'accounts of the nature of matter and its structure' (Weisberg, Needham and Hendry, 2011, p. 46). As an illustration of the shifting sands of the discipline noted above, and perhaps as a start to understanding why many have found it challenging to integrate chemistry tightly into natural theology we might also note that 'Chemistry at any given time is the product of a continuing history, subject both to evolution and on occasion to revolution' (Levere, 2001, p. 17) and 'Complexity, richness, and an economy of means give chemistry its intellectual appeal; utility and application, its universal relevance' (Levere, 2001, p. ix). [Here the tension between Levere's 'complexity' and an 'economy of means' are perhaps reflective of that desire to seek a theology that rejects a rigidity of outcomes and instead celebrates an unpredictable explosion of varieties: one that recognises intense complexity and yet is enabled to build models that in simplicity forge further understandings].

Thus we can see that the uncertainty which informs much of the current quantum mechanically descriptions of atomic and sub-atomic structures is shown to be present in practical ways in chemistry as well. The more mechanistic or positivist philosophical approach that might favour precise outcomes from precise mechanisms utilising precise inputs, simply does not reflect reality. Weisberg, Needham and Hendry propose the epistemic solution of *eliminative induction* to describe this method of reflecting upon mechanisms, where several mechanisms for the pathway achieving a new compound are proposed and as experimentation eliminates certain candidates, the 'probability that one of the remaining mechanisms is correct goes up' (Weisberg, Needham and Hendry, 2011, p. 50). This phrase is however problematic since the preceding discussions indicate that since several mechanisms appear from the evidence to all be 'correct', we should more correctly be speaking of *correct mechanisms*: either the same products are being arrived at via separate routes, or it is not possible to discern which single route is correct, *even if that 'singleness' itself can be proven*. Weisberg, Needham and Hendry also tie-in this epistemic solution to the explanations provided by mechanistic chemists only. This is probably excessively restrictive and any successful solution should in addition encompass theoretical speculations as well, especially given for example the differing ways of representing for benzene (see Appendix A for an introduction to chemical bonding). In this latter case certain ways of representing chemical structures would favour a simplistic approach of alternate single and double bonds between individual carbon atoms in a ring shape, each having a single hydrogen attached:

Figure 1: Benzene

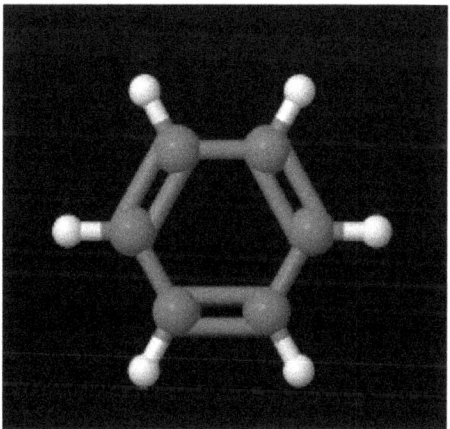

however more recent studies suggest six equal yet non-conforming quasi double-bonds to be more accurate. (Again this latter alters immediately where there are any further groups of atoms appended to the ring). As a result if it could be 'seen', there would be six equal yet 'more-than-single', and 'less-than-double', bonds in a ring. For this reason, 'non-conforming quasi double-bonds' is possibly one explanation of what is apparent in benzene, as opposed to what the chemical physics might indicate: 'hybrid' is another term used for a similar situation where the 'quantum mechanical equations and laboratory chemistry both' (Levere, 2001, p. 179) yield a more nuanced reality: again chemistry appears to be not only crossing boundaries between chemistry, mathematics and physics but now and increasingly over the past decades simply encompassing these other disciplines where such overlap is conducive to a fuller understanding.

This point is illustrated in the volume by Nancy Cartwright entitled 'How the Laws of Physics Lie' (Cartwright, 1983, p. 164) in which she discusses this self-same compound benzene, as well as forms of a further interesting compound dibromobenzene, whose structure may be represented such that the compound could be supposed to exist in two such forms:

Figure 2: 1,4- and 1,2-dibromobenzene

or

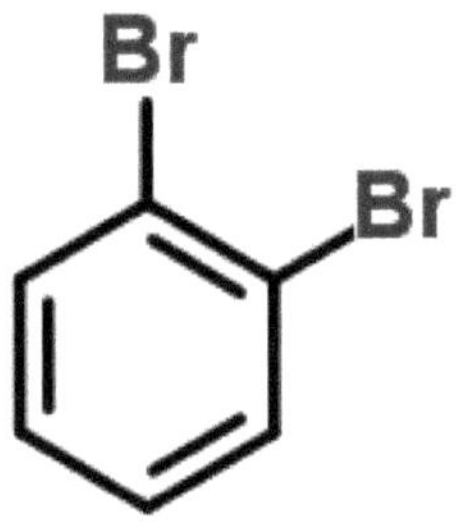

… the first or left-hand diagram representing paradibromo [or 1, 4 dibromo] benzene, and the second righthand image orthodibromo [or 1, 2 dibromo] benzene. However in nature it is known in only one form, which suggests, depending on one's point of view, that it is either both different forms at one-and-the-same-time or that it is one thing inaccurately described. What Cartwright does not say is that the rules used to propose the two alternate structures are used almost universally in chemistry education at various levels to great effect and have enormous predictive power – yet in this case they apparently fail. However in a practical chemical sense, in terms of attempting to understand what one might use the molecule for in some follow-on experiment, the 'erroneous' structure works perfectly well. There is something of a competition between the 'real' quantum mechanical explanation which is exciting and interesting but tells us little and the 'real-world', prosaic, utility-based explanation which with some accuracy postulates the outcome of interactions between dibromobenzene and other reactants: the 'truth' of what is taking place is best described using *multiple* explanations each of which reveals *part* of the truth. Importantly, we have progressed from the situation where it was not possible to discern which single explanation was the whole 'truth', to the deeper truth, whereby a more complete understanding of a particular reaction or process cannot adequately be accommodated within *in a single description,* and instead requires multiple narratives (dare we say 'parables' or 'similies'?) to illustrate all that can be known about it.

Thus to re-use the language employed earlier: epistemic warrant for the particular 'truth' of a proposed mechanism in a chemical process, may often come in the form of models, diagrams and analogies each offering part of the overall 'truth' whose exact nature is unknown and provisional yet where the repeated practice of the process, leads to the same or similar results.

An important aspect of this latter is that by re-utilising the steps of a particular process with similar or related reactants (components of the process), similar products might result. Thus, simply put, if two overall reactions or processes (are ordered or) 'look' the same, there is a degree of epistemic warrant that they will produce similar outputs along similar pathways, this meaning that the truth of the mechanism of a given process might be inferred from similar processes. From this it can be seen that 'similarity' is a useful tool in providing 'scientific knowledge generation' in chemistry (Bengoetxea, Todt and Luján, 2014, pp. 1, 17). Furthermore notions of what 'similar' means are specifically the product of the physical practice of chemistry and not solely, or at times even rather loosely, connected to theory. Yet to what degree is such knowledge derived from 'similarity' indeed 'scientific'? Interesting whilst Bengoetxea, Todt and Luján qualify their words 'knowledge generation' with the word 'scientific', they also refute Quine (Bengoetxea, Todt and Luján, 2014, p. 5) who held that to build knowledge upon 'similarity' was to stand against the logic of mathematics and empirical scientific theory. Yet implicitly Bengoetxea, Todt and Luján's work demonstrates that metaphysical appreciations within chemical research are capable of 'knowledge generation'. As a result, 'knowledge' need not necessarily be qualified by the word 'scientific', in an attempt to legitimise the former. That such metaphysical qualities can contribute to epistemic warrant in the sciences should come as no surprise given the prevalence of the belief in other sciences notably mathematics, that the aesthetic qualities of processes are also indicators of truth. And so it is reasonable to ask why if by way of analogy in chemical processes we have a mechanism for establishing similarity and thereby truth, this truth should be confined to the *physical mechanism* of the process. Surely such similarities could also contribute towards the elucidation of the underlying frameworks at play in the processes themselves? We will return to this in the case of chemistry in the next section, yet for now it is sufficient to appreciate that the process of recognising similarities, is indeed a useful tool in chemical research to discover the truths in underlying processes.

Levere points out in the introduction to his wide ranging volume that the list of known compounds had up to the time of writing (2001) expanded by over seven million in thirty years. At the time of commencing my own MSc studies in the early 1980s, the synthesis of the antibiotic Erythromycin A had just been announced: an extra-ordinary achievement of the step-wise assembly of an organic molecule with twenty-four separate 'junction-boxes' each of which could exist in several different arrangements (or orientations in three-dimensional space), whilst still comprising the same components: yet the target (known) product would only allow for a precise and single arrangement of components around each junction-box. Since that time the effort to create molecules of such – and ever greater – complexity has exploded, now of course often utilising yet another set of differing techniques: genetic manipulation. Taken together chemistry is shown again to be a constantly shifting discipline, with the building blocks, these individual elements known to us, numbering only a little over one hundred and yet 'their possible and actual combinations are so many as to seem infinite' (Levere, 2001, p.ix).

Natural theology has at times been known as a contemplative discipline (Brooke and Cantor, 2000, p. 338) in that it calls upon (Christian) science researchers to evaluate that which is fixed and immutable within the physical realm, as being reflective of the divine agency which lies at its genesis. For chemistry to inform a natural theology, the 'natural' must necessarily come to include not merely the objects of study (which chemistry manufactures itself) yet rather primarily the natural frameworks and mechanisms that give shape to the compounds and processes, so created – that constantly evolving 'continuing history' spoken of above. This must be so, since the 'fixed and immutable' objects (meaning for example cosmic structures and certain laws of physics) which a 'traditional' and contemplative natural theology examine, are as we have seen not fixed and immutable in that same sense in chemistry.

In conclusion then, contemporary chemistry is enormously successful and productive. As to what precisely it is: 'Chemistry is what chemists do' (Levere, 2001, p. 182). Chemical knowledge is gained *a priori* as well as *a posteriori*; art and science: models, graphs, approximations and diagrams might all be blended in its descriptions of the 'why' and 'how' of a given process; the appeal is often more sensual than cerebral (Brooke and Cantor, 2000, p. 314). Chemistry is constantly on the move and yet the building blocks stay the same. Intensely practical, it requires repetition, practice: forms of apprenticeship, before collected insights give rise to further successes. Epistemic warrant for a given underlying mechanism can be provided, utilising the experience of the practical and practiced researcher – experience borne of the knowledge of the context and thus such judgements are context-dependent (Bengoetxea, Todt and Luján, 2014, p. 17) -, by analogy with other similar processes. I have suggested that such analogy should and could not logically be confined to that between similar physical processes, but also to the underlying ontological reasons or frameworks subtending the processes *where such judgements again are being made by those familiar and appreciative of the context*. It will not have escaped the reader that we are on very similar epistemic ground here to that elucidated by Plantinga and others. I have described in chapter 1 above, the importance of the reliabilistic 'cognitive faculties functioning correctly' in the provision of descriptions of 'truth' in the rational holding of Christian belief. We now discover here that there are similar mechanisms being attested to in the provision of 'truth' in the chemical sciences. The types of truth being attested to are both physical as well as metaphysical. They have predictive power and the results obtained thus, are repeatable. We have therefore refuted some commentators' objections to forms of a strict *reliabilism* and expanded it to include not only, as Susan Haack describes it, 'experiential anchoring and explanatory integration of the subject's evidence with respect to a belief' (Haack, 1995, p. 139), but have done so in two different disciplines.

3.2 Chemistry and Metaphysics

3.2.1 Introduction

In the preceding section I have sketched those points in early-modern to contemporary history where chemistry and Christianity have touched. In this section, I explore on what basis contemporary chemical endeavour and Christian theology may be in dialogue together and in so doing discuss the metaphysical within chemistry.

3.2.2 Exploring Critical Realisms

Here I build on a discussion above in which I affirmed that the language our chosen natural theology is described by, must be intelligible to both theologians and scientists. This language supposes a realist outlook on the world where humans are interacting with objects that exist independently of such observers. Such a language will be intelligible to the chemist researcher. Crucially, for the Christian or more generically the religious person Ian Barbour advocates an 'experiential basis [for] religion, which is essential for renewed religious vitality in practice as for a defensible epistemology in theory' (Barbour, 2013, 3064). The Christian should embrace just such a realist outlook, one that also retains a critical perspective in that it is constantly testing and making enquiries of those objects in the world. Thus in this book I advocate a form of *critical realism*. There are differing interpretations of critical realism (McGrath, 2004, pp. 141,143). McGrath is forthright in aligning himself with Roy Bhaskar's interpretation (McGrath, 2004, p. 139). The objective of this section is to explore some of these aspects of 'critical realism(s)'.

My present argument deals in a form of duality in that I assert that there is a qualitative difference between the apprehension mechanisms employed by the Christian and the non-Christian. This will cause difficulties for many: natural theology is about level playing fields in that no one entering into the debate may have recourse to any elements of a revelational tactic: all arguments must proceed purely from the basis of what can be seen, touched, perceived of, the natural world. This is quite correct and both conversationalists are thus on the same 'field'. This is plainly essential in order for Christians to be in relationship with the world they claim to be in love with. As Janet Martin Soskiće remarks:

> We should not let the visionary nature of language [Soskice had earlier quoted from Isaiah 11] distract us from the reality of the call to right relation. The biblical picture is one in which reverence for and right relation with God entail reverence for and right relation with other people made in the image of God, and furthermore right relation

with the rest of the created order. Consideration of our human nature and destiny must be consonant with good science and the understanding it can bring to our biology, our psychology, and our natural genesis. Yet this scientific understanding is framed, for the Christian, by the understanding that we are not just 'natural phenomena', but creatures in the literal sense-we have been created. (Soskice, 2007, pp. 64,65)

If the claims of the life, death and Resurrection of Jesus Christ are to mean anything to those adopted by Him, they must be based in reality, in relationship with the real physical world created through Him, and this relationship must have effect. Having earlier stated that he is 'persuaded of the validity of a carefully nuanced critical realism in both science and theology (2004, p. 10), John Polkinghorne affirms the efficacy of such a reality:

> Belief in the unseen reality of God can properly be defended on the basis of the insightful understanding that it yields in relation to great swathes of spiritual experience, particularly in relation to the gospel record and its testimony of Jesus Christ, and in relation to the continuing worshipful and sacramental experience of the Church. (Polkinghorne, 2004, p. 78)

Thus, the critical realism I am speaking of supports a natural theology that is as has already been said, something entirely normal and normatively intelligible: it is there for all.

The critical realism I advocate here differs from that proposed by Roy Bhaskar, favoured by McGrath, as we shall see. Bhaskar speaks for example of:

> …. the general relativity of our knowledge: viz. that whenever we speak of things or of events etc. in science we must always speak of them and know them under particular descriptions, descriptions which will always be to a greater or lesser extent theoretically determined, which are not neutral reflections of a given world. Epistemological relativism, in this sense, is the handmaiden of ontological realism and must be accepted….. Epistemological relativism insists only upon the impossibility of knowing objects except under particular descriptions…. Philosophers have wanted a theory of truth to provide a criterion or stamp of knowledge. But no such stamp is possible. For the judgement of the truth of a proposition is necessarily intrinsic to the science concerned. There is no way in which we can look at the world and then at a sentence and ask whether they fit. There is just the expression (of the world) in speech (or thought). (Bhaskar,1997, p. 249)

Here, Bhaskar's argument betrays the influence of Wittgenstein's language game theory: truth is relative to the context in which it is promulgated. This makes Bhaskar's thinking something akin to 'epistemological-games' in which truth is relative to the context in which it is promulgated. As for the ways in which such a critical realism might be used to interpret individual reality, Bhaskar remarked earlier:

> If science is to be possible the world must be one of enduring and transfactually active mechanisms; and society must be a structure (or ensemble of powers) irreducible to but present only in the intentional action of men. Science must be conceived as an ongoing social activity; and knowledge as a social product which individuals must reproduce or transform, and which individuals must draw upon to use in their own critical explorations of nature. (Bhaskar, 1997, p. 248)

It is clear from this quotation that according to Bhaskar we can have research and exploration in various disciplines, we can have as he indicates in the same place, 'mechanisms of the production of phenomena in nature', yet these are intelligible only within each of their own respective intellectual domains.

Such a critical realism in a strict sense would not allow me in this present book to be advocating a natural theology which leads the enquirer towards the God whose 'word is truth' (Psalm 119.160; John 17.17). Instead we have as Bhaskar puts it, a dialectic of descriptive and explanatory knowledge, with no foreseeable end. As interpreted by McGrath, Bhaskar proposes the 'critical' faculty enacted by the human as having a layered structure. These layers are the processes the individual performs on their picture of reality as she/he gets to work on the mechanistic, natural, sense-mediated, impulses, of the world we all share (McGrath, 2004, pp. 141–143). Thus Bhaskar's 'transcendental' realism is nothing less than the cumulative effect of the human's filtering mechanism of mind, being brought to bear on that which she/he encounters in the real world. As Bhaskar remarks, 'science as a process is always entirely intrinsic to "thought"' (Bhaskar, 1997, p. 185).

McGrath makes use of Bhaskar's approach to the sciences which refuses to reduce all ultimately to mathematics, which I embrace, although with the insistence that our theological response is *a posteriori,* it would appear to leave little room for faith and a sacramentalism (and with that a clear acknowledgement of the Divine breaking-into the day-to-day experience of humans) which in my view would be a more adequate response to the totality of historical Christian experience (McGrath, 2004, pp. 152–154).

In chapter 2 I described McGrath's approach to Natural Theology as pessimistic and post-fideal, whereas it was necessary for a pre-fideal and optimistic use to be adopted for Natural Theology to function within my proposed conversation. A post-fideal approach is by definition one that is epistemically relativist in a Bhaskarian sense, within that 'fideal' domain. Should we hold strictly to Bhaskar's approach to critical realism then a mutually intelligible conversation between differing intellectual domains or disciplines becomes difficult and the thesis of this book would likewise be in doubt. Thus I require a critically realist approach but not a Bhaskarian one.

Within this book wherever the processes in chemistry are described it will become obvious to the reader that models and modelling play an important role and I draw to the readers' attention variously, diagrammatic representations in chapter 3 and the role of mathematical modelling mentioned in the same place as well as the explanations offered in Appendices A and B. A given complex issue is not being *reduced* to a model or models, yet differing aspects of an overall truth are being individually explored through the use of these methods and then finally the whole is being reviewed and re-built within the person's understanding to provide a richly layered tapestry of comprehension. Again such a degree of comprehension of this single complex issue is then open to revision when newer or improved models become available: the complexities are being constantly *critically* appraised in the light of continuing exploration. The reader can see such methodologies at work in the review of papers in chapter 4.

Yet such ways of coming to an improved understanding of truths in science are not restricted to the sciences alone. In religious discourse Barbour says 'Inherited models are for many individuals today almost totally detached from human life. The experiences which traditional models once interpreted are in large measure ignored or suppressed' (Barbour, 2013, 3068) although crucially all that is being suggested by this is that the older models are not appropriate for their contemporary task and audience: the experiences themselves are not being disputed. Taking his brief from the theological doctrine of the Trinity, Polkinghorne says somewhat amusingly 'Much of the writing on the Trinity is formidably technical in its character …. people speak of, begetting and procession, of foliation and aspiration, of perichoresis and appropriation, …. (at) times one is driven to wonder "How do they know?"' (Polkinghorne, 2004, p. 91). He then later in the same place suggests that both scientists and theologians at times make assertions beyond what our experience might allow for: we have ceased to be *critical* in our *realism*. And yet Polkinghorne earlier is firm in stating that he is wedded to a form of critical realism whilst not wishing to move away from the 'grand scheme of Trinitarian theology' (Polkinghorne, 2004, p 10).

Thus we need an improved or altered set of models in order to achieve as Barbour says, 'a greater awareness of the experiential correlates of theological concepts' (Barbour 2013, 3071). From a religious perspective, the experience for example of the beauty of God could likewise be expressed as the experiential correlate of the theological concept of the beauty of God. In this way 'the idea of models in the interpretation of such experiences may answer some of the difficulties in talking about God which are now felt so widely' (Barbour 2013, 3079). As Soskice remarks:

> ... in recent decades some theologians have regarded science not as an enemy but as an ally, and have called attention to strategies of scientific theory construction and model building in defence of their own strategies of theory construction and model building. As modern science has become more eloquent about its own limitations and the difficulty and tentativeness of any truth-claims, theologians have been emboldened to make comparisons with their own tasks. (Soskice, 2007, p. 54)

Much of what the current book explores in terms of beauty as perceived within chemistry amounts to a re-drawing of such models for a contemporary audience. As will be seen below, in contrast to the beautiful art of old which can be seen and touched, chemical compounds and processes which can not be seen are being described as being beautiful by chemists themselves. Suddenly we have a language – that of beauty – which itself becomes something mutually comprehensible between science and religion or more accurately in this present book, between Christianity and chemistry. And at all times the position being adopted is critically realist.

(I have already admitted above, that the theologian in the conversation is holding to a firm Christian Faith in their personal practice whilst conversing in terms devoid of revelation, when speaking with their chemist partner. Let the reader note at this point the distinction between acknowledging that there is a God and the fact of being a Christian. Christian Scripture is clear that it is possible for a being to know that there is a God, but to reject the Christian Faith (James 2.19). Where I use such terms as pre and post fideal, I am using them in respect of the Christian Faith and not in respect of an acknowledgement of there being a God. Thus a partial objective of the conversation proposed in this book is that the chemist might come to understand that it is entirely rational – Barbour's 'defensible epistemology' – to accept that there is a God, whilst at the same time not coming to a place of Faith in the Christian God on the basis of such reasoned arguments alone).

3.2.3 Christ Perceived in Creation

If chemistry is to inform a Christian natural theology, then it is necessary to show that any aesthetic response arising from either theology or (chemical) science, emanates from an encounter with the Christian God, and is not merely a 'theistic' response, in the sense of an anonymous encounter with a nameless Numinous or Transcendent 'other'. Thus together with Polkinghorne I make a plea here for a natural theology that is not merely theistic but indeed Christian. He speaks in a memorable phrase of Einstein's attitude towards God that 'this attitude will not do for Christianity', and that the 'Christian God is the Ground of the

hope of a destiny beyond death, both for human individuals and for the cosmos itself'. He, again memorably, rejects the 'semantic plasticity' found in the writings of certain contemporary writers (Polkinghorne, 2004, pp. 93–96). Instead he advocates an overall theistic approach 'within the Christian context' that draws 'inspiration from the Bible, and in particular from the life and words of Jesus of Nazareth' (Polkinghorne, 2004, p. 16): it must plainly be 'of Christ'.
Thus it is necessary to be clear who the Christ is:

> Do not let your hearts be distressed. You believe in God; believe also in me. There are many dwelling places in my Father's house. Otherwise, I would have told you, because I am going away to make ready a place for you. And if I go and make ready a place for you, I will come again and take you to be with me, so that where I am you may be too. And you know the way where I am going." Thomas said, "Lord, we don't know where you are going. How can we know the way?" Jesus replied, "I am the way, and the truth, and the life. No one comes to the Father except through me. If you have known me, you will know my Father too. And from now on you do know him and have seen him." Philip said, "Lord, show us the Father, and we will be content." Jesus replied, "Have I been with you for so long, and you have not known me, Philip? The person who has seen me has seen the Father! How can you say, 'Show us the Father'? Do you not believe that I am in the Father, and the Father is in me? (John 14.1–10a)

Hence from this passage and others we assert that Christ is God, yet quite plainly and again from this passage, is distinct from God known as 'Father'. Elsewhere in scripture we know of God as 'Spirit'. From this it is plain that there is a 'three-ness' within God, a 'Trinitarian' reality of God. And again, it follows that it is precisely the reality of the Trinity that must be being witnessed in any encounter in the natural world, if I contend that this world was created by God, which I do. Thus it is necessary to digress into Trinitarian theology, if only briefly, to explore the nature of the aesthetic experience that the scientist is seeing in their investigations. Having justified a natural theology theologically in Chapter 2, here I want to demonstrate what it is about specifically the Christian Trinitarian God that is being encountered in the natural world by those scientists who strive in this field. For this it will be necessary to move back into Christian revealed 'God-talk' to show the source of that which is revealed in the natural world.

Christ is variously described in the Christian New Testament writings, as the One through whom all of creation emanates (Colossians 1.15–17), and together with this the One through-whom God created the eons or worlds – having the sense of creating both substance and time (see Hebrews 1.1–2).

The manner in which the understanding of God as being Trinitarian came about, requires some explanation. At many places in the Hebrew Scriptures the Spirit of God is mentioned, yet in no place do we have a firm interpretation of the

Spirit being of a differing personality or aspect of the Divinity. The Christian or New Testament scriptures, most especially in John's Gospel yet elsewhere as well, declare the divinity of Christ. They also provide ample proof that this is what they knew to be asserting: the writers appreciated what their words amounted to (see for example John 20.31). Such proofs would appear to have been provided for those well versed in the Hebrew Scriptures such that they show Christ performing actions that could only have been performed by someone who was God, in the understanding of such people. The effect was to create something of a happy confusion in that the early Jewish converts to the Christian religion, being fiercely monotheistic, were forced by such assertions to confront the reality that God was in heaven, and God was standing in front of them. Christ answered such confusion by stating that there was no difference in motivation and essence between the One He called His Father, and Himself. Furthermore Christ was careful to state that there was a hierarchy and that the Father was greater than Himself.

In addition Christ spoke of the Spirit in new ways. He gave the Spirit's Name, 'Paraclete' and told of how there would be a sharing of something of Himself by the Spirit to his followers, the opening event of which Christian history knows as Pentecost (Acts 2). This afforded great power to the early Christian followers and caused considerable numbers of people to 'be added to the number of' disciples of Christ.

As a result the early Christians witnessed absolutely to a monotheistic view of God and yet they were witnesses to three manifestations or persons of God, and indeed within God. Over the early centuries this witness became the doctrine of the Trinity as set forth in the pronouncement of councils of the early church. Such a realisation of relationships within God, naturally gave rise to thoughts of what the nature of these might be. Immediately it became clear that the Father and Son had a relationship of intense love shared with the Spirit. When questions arose as to how the Father could allow for His Son to experience such suffering in death, prior to His resurrection, parallels were drawn for example with the ancient account of Abraham and his son Isaac (Genesis 22) and the rightful treatment of a son by his father (Hebrews 5.7–10).

The words of Proverbs 8 state:

> I love those who love me, and those who seek me find me. Riches and honour are with me, long-lasting wealth and righteousness. My fruit is better than the purest gold, and what I produce is better than choice silver. I walk in the path of righteousness, in the pathway of justice, that I may cause those who love me to inherit wealth, and that I may fill their treasuries. The Lord created me as the beginning of his works, before his deeds of long ago. From eternity I was appointed, from the beginning, from before the world existed. When there were no deep oceans I was born, when there were no springs

overflowing with water; before the mountains were set in place – before the hills – I was born, before he made the earth and its fields, or the beginning of the dust of the world. When he established the heavens, I was there; when he marked out the horizon over the face of the deep, when he established the clouds above, when the fountains of the deep grew strong, when he gave the sea his decree that the waters should not pass over his command, when he marked out the foundations of the earth, then I was beside him as a master craftsman, and I was his delight day by day, rejoicing before him at all times, rejoicing in the habitable part of his earth, and delighting in its people. (Proverbs 8.17–31)

This passage when viewed in context, is speaking of the wisdom of God present at and in, the Genesis creation act(s). The point has already been made (in the context of this book) that a Christian view of these matters must be 'of Christ' and reflective of Him – simply a theistic view will not do. Since Christ is known in Christian tradition as both the power and wisdom of God (1 Corinthians 1.24), it is a simple matter to identify the properties of the wisdom of God spoken about here in Proverbs – that sense of extravagant joy and delight in admiring the creative act – as being signatures of the Christ through whom the Universe was created. By way of illustration of this point see for example Enid Mellor (Mellor, 1999, pp. 60, 61) who comments on how the 'new testament writers are using Proverbs 8 to fill out what they want to say about Jesus' and also J D Martin who is explicit that 'many … regard the presentation of wisdom in Proverbs … as an hypostatis, an aspect of the Godhead', (Martin, 1995, pp. 87, 88). Furthermore, for where what is said of wisdom is being conflated with both the promised messianic king and with God, see Brown, Fitzmyer and Murphy, 1990, p. 457. Yet from Genesis 1 it is plain that it was the Spirit present also at Creation. This apparent tension between the Son and the Spirit as to which Person of the Trinity we should be referring to, had already been resolved and indeed dissolved much earlier by one of the Church Fathers, Irenaeus, who in his Book IV, chapter 20.3, of *Against the Heresies* remarks:

> I have also largely demonstrated, that the Word, namely the Son, was always with the Father; and that Wisdom also, which is the Spirit, was present with Him, anterior to all creation, He declares by Solomon: 'God by Wisdom founded the earth, and by understanding hath He established the heaven. By His knowledge the depths burst forth, and the clouds dropped down the dew.' (Irenaeus, 2015)

He then goes on to quote parts of the same passage from Proverbs 8 given above. Earlier in this same chapter 20 of his book IV Irenaeus, having explicitly identified the Word made flesh with the Christ and the Spirit with Wisdom, explains that the same God who 'made all things by the Word …. adorned them by [His] Wisdom'. And so we have the sense of the power of creating aspects of the physical world exercised by the Word and the power of communicating their properties being exercised by the Spirit. This accords well with Christ's own explanation of the actions of the Spirit when he states:

> But when he, the Spirit of truth, comes, he will guide you into all truth. For he will not speak on his own authority, but will speak whatever he hears, and will tell you what is to come. He will glorify me, because he will receive from me what is mine and will tell it to you. Everything that the Father has is mine; that is why I said the Spirit will receive from me what is mine and will tell it to you. (John 16.13–15)

God is being shown to act and then within God-self to communicate those actions; the Word in the flesh performed actions at Creation, performed actions whilst amongst us and then in history these truths are communicated by the Spirit. In fact this action of communication goes far beyond mere speech as we might see it. In this way the Wisdom of God is seen to be communicating the character of God as manifest in the actions of the Word of God in Creation, which are precisely these perceptions of glory, of wonder and of beauty when confronted by the physical or natural world. These perceptions far exceed mere 'feelings': they are the gifting of the power to act on account of the realisation of whom it is that gave rise to these perceptions: God is revealed through 'His two "hands", the Son and the Spirit, [….] who made the world through them' (Sherry, 2002, p. 4). It is worth noting that Jesus as a human being was always God from the moment of the Incarnation yet was only revealed and communicated to those around Him as the Christ, the Messiah, after the Spirit came upon Him at his baptism. It was after this event that the Gospels start to record the 'signs and wonders'. It can be seen that God in Creation is revealing himself to people and becoming active as Cause, Maker and Perfecter (Sherry, 2002, p. 5 after Gregory of Nazianzus). Perhaps many could make an artefact but only a 'master craftsman' can 'perfect' and 'adorn' it.

I note the words of John's Gospel 1.17–18: 'For the law was given through Moses, but grace and truth came about through Jesus Christ. No one has ever seen God. The only one, himself God, who is in the bosom of the Father, has made him known'. From this it is plain that a Spirit-communicated and Christ-centric view of action in the creative sphere, is quite naturally accompanied by feelings of joy, of delight towards the One who created them, of a sense of artistry, of a delight in and with those in the world, and all of this is clear from the Proverbs 8 quotation. I am speaking here of the reaction of observers, people, to aspects of the creative power being exercised within the natural world. The descriptions here are quite distinct from for example the words of Psalm 19 which speaks of the grandeur of the created world. Now here in contrast we are naming the agency through whom these very artefacts over which we rejoice, were and are created: there is every sense in which these created entities are reflective of the 'master craftsman' who gave birth to them. Thus *the created artefacts give rise to a sense of joy* simply on account of the craftsmanship, the mark of the Maker, of the artefacts so created. And as we have already pointed out, this Person is the one who 'is love' (1 John 4.16) and is in this Trinitarian relationship of love, within God Himself. And since it is Christ who has 'made him known', it is *an attribute of Christ* as communicated to us by the Spirit, that any sense of wonder or delight at aspects of the created world and its creative powers, is witnessed to, *when those in it experience such appreciations*. It is a reaction borne quite naturally out of the nature of whom God is, that both God Himself and those He has created, should experience a sense of joy, of marvelling at, the craftsmanship of the artefacts in the created world, marvelling at the context of the 'habitable earth' which sustains these natural wonders. These natural wonders are many and various and include those seen in chemistry research. It is important to note that this conclusion has been drawn both when seen from a theological perspective as in the brief exegesis of the Proverbs *pericope* above, as well as from the reactions of scientists themselves (as noted in the work of Derkse above). There is a distinct *separation* between the Creator and created, yet we witness to a *communication* such that the source of Creation is being alluded to within that which is created.

3.2.4 Which Critical Realism?

In conclusion, I adopt a version of critical realism in response to the created world. It is 'realistic' in that it involves sensory interaction of and with the real world, an objectivity outside of oneself. It persists in being 'critical' in that this form of realism retains a sense of the contingent. The chosen approach to critical realism must allow for a causal link, however loosely revealed, between the God who created the Universe and the created artefacts (and their beauty) which show forth that creative action. There is therefore a rational explanation for the arising of a perception of beauty and this rational explanation amounts to a cause for the arising of such perceptions. Bhaskar explicitly rejects the existence of any such form of 'ontologically basic' explanation (Bhaskar, 1997, pp. 49,50), which plainly Polkinghorne, Soskice and others affirm, and appeals instead in the same place to 'generative mechanisms of nature'. From this it appears reasonable to suggest a difficulty at the heart of Bhaskar's argument since if nature can support a fixed 'generative mechanism' surely this amounts to an 'ontologically basic' explanation for whatever event is being described? In addition, surely a God who created the Universe could plausibly put in place just such mechanisms? And these perceptions are not epistemically relativistic precisely because they can be perceived by persons working in differing disciplines.

The Christian is enlightened by the example of Christ and is thus emboldened to serve their neighbour in the messiness of the everyday. Their unfolding and at times unpredictable experience of the natural world constantly leaves them aghast at the new possibilities and indeed realities of creation. For both the Christian and the non-Christian such an unfolding experience includes any practice of the scientific method in realising these new possibilities, possibilities that are themselves falsifiable:

> .... why go to the trouble involved in doing science if one does not believe that thereby we are learning what the physical world is actually like? If you take this realist view, then unpredictabilities will not be seen as unfortunate epistemological deficits but rather as signs of an actual ontological openness to the future. (Polkinghorne, 2004, p. 79)

This sense of humility in the face of the grandeur and mystery of what God has created – and is redeeming! – in Christ, should serve us well. Our minds, constantly in hope seeking God's Spirit in guidance, our forever modelling and remodelling our responses. As a result and as Ian Barbour remarks, 'Christianity should never be equated with any metaphysical system', there being dangers in seeking to box God into our own view of things (Barbour, 1990, p. 30). For this reason it would appear to be entirely valid to abstract and re-use aspects of differing models, in a continuing quest to seek to answer that which 'burns within us' (see Jeremiah 20.9 and Luke 24.32) whilst insisting that it is God who put the Universe in place. Barbour speaks in the same place of taking a position of 'independence' within an overall embracing attitude of dialogue when discussing the place of humanity within creation, and this would appear to be the most sensible form of realism; truly resting upon an evidence-based system of exploration.

3.2.5 The Aesthetics of Chemistry

Within this present section we are attempting to delineate the field within which meaningful discourse may be held between natural scientists and theologians. It appears that the 'meta' language of science as expounded in the Philosophy of Science may be most fruitful.

McGrath argues:

> the Christian understanding of the ontology of creation demands a faithful investigation of nature. For this reason, the exploration of the interface between Christian theology and the natural sciences is to be regarded as ontologically motivated and legitimated. Yet the Christian doctrine of creation is not limited by the demand that we see nature as creation; it has a highly significant Christological component. (McGrath, 2001, p. 24)

In a section toward the end of this part of the project we shall return to this point that McGrath raises: where in our appreciation of the significance of nature as observed does our gaze turn from nature to the centrality of Jesus? And can we agree with McGrath that meaningful discourse between Christianity and the natural sciences holds significance only in terms of the nature of our existing? Is it not rather to do with an encounter between an individual and the person of Jesus? Indeed, is not our understanding of aesthetics shaped by an encounter with God in Christ (Psalm 27.4, Song of Solomon 2.1, Matthew 6.28, 29)?

Before coming to any conclusions in this section it should be noted that McGrath makes much of the appreciation of aesthetics (McGrath, 2001, pp. 234–240) saying for example:

> A strong doctrine of creation (such as that associated with Christianity) leads to the expectation of a fundamental convergence of truth and beauty in the investigation and explanation of the world, precisely on account of the grounding of that world in the nature of God. (McGrath, 2001, p. 240)

This author would agree in large measure. However he then goes on in the same place to conflate the 'nature of the Creator' with the 'ordering and regularity of creation'. We seem to have arrived at an appreciation of aesthetics which has to do with this same 'ordering and regularity'. Surely if beauty is solely or largely about order and regularity it would be a poor subject of study indeed.

Furthermore Alan Padgett rightly cautions that not all appreciation of beauty is the same:

> What counts as a beautiful solution in mathematics is one guide to a good answer to a problem, yet what counts as a beautiful solution in chemistry is quite different. In both a kind of rational elegance is seen as a kind of guide to a good answer, but what counts as elegant in these disciplines is distinct. (Padgett, 2012, p. 94)

Perhaps the following quotation from Roald Hoffmann will illustrate to the reader the way in which the aesthetic ideal in chemistry truly illustrates the arresting manner in which the human mind can transcend the physical and apprehend beauty in a remarkable and truly metaphysical sense, whilst being rooted in the physical world:

> I recently saw a beautiful molecule in the literature. The authors, E. Nakamura and collaborators (Nakamura, E.; Tahara, K.; Matsuo, Y.; Sawamura, M.: 2003, 'Synthesis, Structure, and Aromaticity of a Hoop- Shaped Cyclic Benzenoid [10]Cyclophenacene', Journal of the American Chemical Society, 125 (10), 2834–35), explained by way of introduction how they and others had long sought to make 'hoop' type compounds, in which there is delocalization of electrons around a finite cylinder. Note the muddle of compounds (materials), molecules, and models in my, and their, language. (Hoffmann, 2003, p. 3)

Here beauty is said to be being represented in an entire confusion of attributes of the process: there is the physical that can be seen (the compounds and their physical presence), in the discreet output or product as a molecule (which can not be seen and only inferred from the sampling apparatus), in the 'paper' representations of that product (where the diagrams may, as is often the case, be drawn such that they are 'pretty' and not) and finally in the language used to describe the result. The aesthetic ideal in chemistry is plainly borne out of the training, experience and contextual appreciation (the history of achievement to date and the professions' hopes for the future) of the practitioners. There is an attitude which speaks of an ability to appreciate the wonderment as perceived.

From what does this 'attitude' arise? How might the didactic process of acquiring scientific knowledge, sufficient for a lifetime of service, be described? Padgett is right to compare two disciplines with distinctive rules and traits, and into which researchers are as-it-were inducted as apprentices:

> To learn a natural or human science is not simply to be trained in pure a priori logical reasoning or in universal axiomatic systems of deductive truth, but is closer to being apprenticed into a valuable skill which requires mentoring into a community of experts, a way of thinking, an angle of vision, and a specific labor. A student of any specialized science is thus inducted into a community of truth-seeking fellow scientists, whose reasoning is shaped by that tradition of inquiry ….. The epistemological values which are embedded in the contingent, historical, and humanly constructed sciences (academic disciplines) are not pure noetic truths – at least, most of them are not – yet with successful and fruitful sciences they should be given prima facie epistemic warrant unless there is some reason to doubt them. So the specific sciences are best understood as using both formal and informal logics. (Padgett, 2012, p. 94)

The point being that it is as the human researcher plays with, elaborates upon, builds upon, those rules which have shaped their training in chemistry, creating and forming new methodologies, processes and products – it is under such circumstances that beauty is perceived: very much in the practical 'doing' of it. Beauty can be *shown* (to the individual, to you) by others and then experienced oneself, yet it transcends explanation (Derkse, 1997, p. 48).

Chemistry has already been described as breaking down the barriers between nature and art (Brooke and Cantor, 2000, pp. 315, 319, 329) and although the authors speak here of chemists 'defying' nature, elsewhere through Process theologies as discussed below, it could even be suggested that it is 'deifying' nature. Yet this discussion of defying and deification can be set aside if we start to see beauty as using our God-given faculties to interact with what God in Christ has provided: to create and renew, to interpret and codify, through a vision of this 'beauty' as our guide. Now there is no longer the need to distinguish nature and artifice in scientific (chemical) research: the prejudice against chemistry as being 'unnatural', is set aside.

And it is precisely an envisioning of this aesthetic sense and sensing, when we employ the *language* of beauty, which is of importance. It is not that we should all agree precisely on a particular aesthetic standard, or that the thing which is aesthetically pleasing in the arts has the same qualitative value as something perceived in the sciences. Indeed Schummer in part suggests that this is a lost cause (Schummer, 2003, p. 98). No, it is rather that there is a 'something' which participants can declare exhibits *similarity* between the artistic and natural scientific disciplines, which we can agree to call 'beauty' and which is not (simply) a feeling or an emotion, yet rather possesses true value and points to something beyond that which can be measured. (And of course we are using 'similarity' in the same sense as used in the preceding section). At times this sense of true value is expressed as that aspect of beauty which inspires and guides innovation, that is, the origin for new ideas (Schummer, 2003, p. 99): it is thus in that sense real and concrete. In fact we must move beyond that sense of the beautiful which merely satisfies a need to see something pretty, to that sense of the aesthetic ideal which drives us toward ever greater achievement, an ever greater encounter with the thrilling truth of the real. This Schummer expresses in the same place as the 'aesthetic experience [becoming] part or even a driving force of the research process'. Here the 'research process' is human endeavour expressed in chemical explorations, yet who would or could disagree that it is a similar desire to see excellence in the aesthetic realm which drives artists of all descriptions towards ever greater creations? (It is helpful to note that any response to perceived aesthetic excellence can be both subjectively and culturally mediated, demonstrating that our sense experience is the mechanism and the person is the interpreter, leaving the notion of the perception of beauty as aesthetic excellence, intact – Schummer, 2003, pp. 81, 82). Once again it appears entirely rational to attest to the concrete reality and utility of the aesthetic ideal of the beautiful.

Crucially for our present study, Schummer was attempting in this noteworthy paper to 'investigate if certain parts of chemistry are comparable to the fine arts' (Schummer, 2003, p. 74) and it should not therefore, given that as he says the 'theories of art are rather divergent and frequently in too bad a shape for that purpose', surprise us when he says that 'chemists' claim to the beauty of certain molecules cannot be justified, because every attempt at developing an aesthetics of molecules finishes up in a blind alley' (Schummer, 2003, pp. 97–98). It is questionable to declare something to be impossible simply because it has not yet been done. He states on the following page, and I have already alluded to this above, that aesthetic experience guides and inspires innovative ideas in chemical research and that future work should explore 'where and how aesthetic experience becomes part or even a driving force of the research process' (Schummer, 2003, p. 99). Plainly it is logically impossible for something to *not* exist and also to be a 'driving force' in something. Thus in summation of Schummer's argument, beauty in chemistry exists and is a 'driving force', we simply don't know why or indeed how, if we go searching for an explanation in the fine arts. Furthermore whatever type of beauty we are talking about that is so fundamentally important that it orchestrates chemical research, it is not the same type as that seen in a fine painting or sculpture, but utterly crucially *it is beauty and the chemistry is beautiful*. (In passing it should be noted that this form of the beautiful exists in other more modern disciplines, including software design. Padgett above also speaks of beauty in mathematics and there is I would suggest even less likelihood of establishing a connection between mathematics and the fine arts on the basis attempted by Schummer).

And what of the researcher, now as we can see so often driven by a sense of the beautiful, what of their response to that experiencing of the non- or un- scientific (if ignorance of the how and why can be termed 'unscientific') within themselves, as giving them that real urge to tap the aesthetic in their work? Frequently they respond with gratitude, respect, reverence and even adoration (Derkse, 1997, p. 52) and as Derkse goes on to say in the same place, such responses can only be given meaning, can only be correctly understood, when seen in the light of those transcendental experiences of religious believers. These and other such 'motivators' come from a joy in revelling at one's connectedness to those one cares about and loves. Not only is scientific language redundant at this point, it becomes subsumed into the exchange between God and those He has created as being reflective of all that we have been given to 'rule and subdue' (Genesis 1.28), indeed all that is there and that He has put there, for us to recognise Him by. Derkse does not go quite so far, however such a relational interaction between as he puts it, 'person and Person' may now be shown to be what is at work in such reactions to scientific exploration and achievement. Derkse himself gives some weight to such conclusions through the conflating of these responses (that is the responses being described here, of perceptions of the beautiful) to those experienced in for example musical appreciation.

In conclusion on this point we appear to have determined that the language of aesthetics, of wonder, pattern, form and simplicity (used within models to reduce complexity), of elegance [in the sense of the *art* of making a good choice (Derkse, 1997, p. 49)]: all these are seen in both theology and science. If we are to achieve any form of common language or understanding between theology and chemistry, it would seem that we can look here. More than this, chemistry as a scientific discipline is *the* discipline which powerfully demonstrates this ability to bridge any perceived gap; again it is *the scientific* discipline where-in that which is beautiful may often with ease, be apprehended. Yet caution must be exercised in transplanting the exact notions of aesthetic quality from one discipline to another – there is context-dependency to consider as mentioned elsewhere. In this book I appeal to the human *capacity* to discern aesthetic quality in the practice of any given discipline, as a wondrous attribute. Derkse for example speaks of scientists *experiencing* certain traits as being 'aspects of beauty' (Derkse, 1997, p. 48). This I assert is itself indicative of a metaphysical quality to human thought: humans can envision beyond that which is perceived. In addition I remind the reader that only certain aspects of contemporary chemistry are being surveyed. It is a very wide field of endeavour and not all aspects may equally or at all, show aspects of beauty to the enquirer.

Yet even if the scientist, our 'honest enquirer', is enabled to discern a numinous or even a transcendent quality to perceptions of their work and exploration in the chemical sciences (the beginnings of a natural theology), they have little reason to suppose that such attributes come about as a result of the Christian God being demonstrated through such a natural theology. In this conversation between the natural theologian informed by chemistry, and their chemical scientist, how do they point the way to the Christian God? What forms of theology might be suitable? In what follows I will discuss the suitability of one candidate theology that has at times been implied: Brooke and Cantor speak of 'a kind of process theology' (2000, p. 315). From this it is clear that an aspect of the Whiteheadean scheme is in view given that it lies at the root of such thinking.

3.2.6 Process Theology: a Child of Certain Strands of Process Thought

In the discussion of natural theologies above, the components of such a theology that were commensurate with a rational epistemology, were set out. Natural theologies are part of the discipline of Systematics. A natural theology on its own, being as it is separate from revealed theologies, is not a complete systematic theology. A natural theology, if it is to be part of an holistic understanding of the Christian God must demonstrate a coherence with and a context within, such a more complete systematic theology.

Amongst those authors where Chemistry has been explored, aspects of certain Process Theologies have most frequently been used to underpin the connections between Chemistry and natural theology. A brief exploration of what these Process Theologies are will now be embarked upon. There remains the whole question of the underpinnings of Process metaphysics and its suitability for establishing a coherent and meaningful view of reality. There is in this manner a considerable separation between the fundamental initial thought and the way that thought has been developed into various stands of metaphysics. These will be dealt with later.

The beginnings of the philosophies associated with such theologies are most often associated with Alfred North Whitehead whose Gifford lectures of 1927–28 will form the basis of our initial understandings. Whilst I have used the term 'a systematic theology' as the generic term for an individual coherent set of understandings of the Christian God which would then underpin a strategy for drawing the Chemical sciences into a partnership with natural theology, Whitehead uses 'speculative philosophy' as the 'necessary system of general ideas in terms of which every element of our experience can be interpreted' (Whitehead, 1978, p. 3). This definition is at least helpful in that it is analogous to our own search for a systematic theology to underpin our natural theology.

Process theologies deriving from this philosophy have developed considerably from their early beginnings making precise judgements about what contemporary philosophers and theologians mean, difficult. Donald Viney for example holds that Whitehead believed in the objective reality of God (Viney, 2014, p. 4), but given what follows it will be seen that this is in no way clear and certainly modern followers of this movement would certainly not believe in God's concrete existence. As a result it will be necessary to collect certain common understandings or traits rather than expecting to find exact or precise definitions common to all Process thinking.

The first and most important of these comes from Whitehead himself:

> the way in which the contemporary actual entities are relevant to the 'formal' existence of the subject in question is the first example of the general principle, that objectification relegates into irrelevance, or into a subordinate relevance, the full constitution of the objectified entity. Some real component in the objectified entity assumes the role of being how that particular entity is a datum in the experience of the subject. In this case, the objectified contemporaries are only directly relevant to the subject in their character of arising from a datum which is an extensive continuum. (Whitehead,1978, p. 62)

Thus Whitehead contends that the truth or reality of something has to do with its 'becoming' within this 'extensive continuum', rather than any static ontological identity: 'each actual entity is itself only describable as an organic process' (Whitehead, 1978, p. 215). A contemporary commentator, Rod Garner, puts it differently but says much the same: 'change defines the surface of things, however much we crave the consistencies of habit and routine' (Garner, 2011, p. 86). Furthermore 'god' is often in Process thought also included as one of these forever-changing 'somethings', and in consequence there can be no 'god' that is an objective reality or person or God outside of this temporal sphere. As Viney remarks:

> Implicit in traditional theism's doctrine of creation are the ideas that God's creative act and God's knowledge of the world are non-temporal …. Process theism takes a contrary view that time is the process of creation. (Viney, 2014, p. 4)

In some Process thought 'god' is allowed to remain as an objective reality with which humans work in partnership to realise the constant development of the physical realm, and certainly Brooke and Cantor describe Priestley as thinking in this manner (Brooke and Cantor, 2000, pp. 326–329). Thus the notion of salvation in the sense of human progress and flourishing is perceived as being about humans working together with God in partnership: 'Redemption, understood as the duty and joy of holding the world dear, is something we undertake in partnership with God, and all friends of God' (Garner, 2011, pp. 95–96). In fact Whitehead went so far as to say 'It is as true to say that God creates the World, as that the World creates God' (Whitehead, 1978, p. 348), which I imagine is an expression of some ultimate form of partnership. It is this aspect of partnership which is also a common trait in Process thought.

The reader will immediately notice a confusion in precisely what a 'god' or 'God' is in this field. Is God an objective reality or a construct of the mind welling-up into a society (see Whitehead, 1978, pp. 34–35); does God uphold the Universe He created or is He a fellow-traveller within this physical world or again some construct of the human mind and society? When Johanna Seibt (2013, p 1) describes Process philosophies in terms of the 'role of mind in our experience of reality as becoming' we start to understand that these ways of understanding the Universe around us have to do with objectifying the reality of collective human consciousness. In this way the only things that are apparently fixed are in fact 'dynamic organisations that arise due to the continuously ongoing interaction of processes' (Seibt, 2013, p. 1), and societies are one such.

It is plain why theologies deriving from Process philosophies might be thought useful in 'exploring the space between chemistry and natural theology' (Brooke and Cantor, 2000, p. 315): chemistry might indeed be seen as a partnership between the creative divine and humans creatively working with the tools the divine has provided. Furthermore in perceiving God as part of this physical world, revelatory theologies are circumvented (Viney, 2014, p. 53) and there is no supernatural intervention into this natural world (Viney, 2014, p. 6), thus fitting the definition of a natural theology. Such theologies however are never far from considerations of theodicy and the possibility of God: once chemistry can be seen to improve matters, questions of the harms it may also cause and the potential vulnerabilities of a God who allows humans to work in partnership with Him, are more readily entertained.

In this regard, Arthur Peacocke remarks:

> in recent years process theologians have extensively used the analogy in the form self:body::God:world ….. [Where] holistic conceptions of the human person which have arisen in response to scientific knowledge and which conceive the 'self' and the 'body' as two aspects of one total unity, then serve in combination with this analogy to facilitate a non-interventionist way of thinking of God's agency in the world …. (Peacocke, 1993, p. 167)

Hence a knowledge of God *in the world* is perceived *by action through an agency*. Later Peacocke goes on to state that whilst God may validly be perceived as the 'ultimate Reality', God also gives this being 'to all other, dependent realities that constitute the world' and that in this sense God can not be unchanged through such a sharing and thus that the 'dynamic divine' is both becoming as well as being (Peacocke, 1993, p. 184). In terms of an understanding of the nature of God as explained through the Christian scriptures, it again would be challenging to think of God as 'becoming' when Christ is described as being the same 'yesterday, today and forever' (Hebrews 13.8). Yet Peacocke later says that such an understanding of the 'becoming God', provides the basis for the possibility and indeed hope that 'the immanence of God in the world might display, in humanity at least, a hint of, some kind of reflection of, the transcendence/immanence of God' (Peacocke, 1993, p. 187). However from this, it is once again more difficult it would seem to sustain a natural theology that might speak of *revealing* or *hinting at,* the underlying presence of specifically the Christian God upholding and sustaining the Universe, having created it *ex nihilo*. Peacocke goes on to deny the physical resurrection of Christ (Peacocke, 1993, p. 281) since this is inconsistent with our understanding of what happens physiologically at death. The ancients were not ignorant of such detail (John 11.39) and yet on the basis of the evidence before them, chose to believe. Thus the language employed here of a Process Theology is not sustaining of an orthodox Christian faith: as Polkinghorne remarks 'the process God does not seem to be the One who could have raised Jesus from the dead' (Polkinghorne, 2004, p. 19). Process theology does not provide a bridge as utilised in this specific instance, between the natural sciences and the Christian Faith. It is important however to note that Peacocke is attempting to navigate a way through competing visions of God where the one suggests God as invariant, unchanging Creator, infinitely good and as such incapable of walking and being alongside humanity, and the other a God who is touched by and through human suffering yet in consequence (according to the argument) is incapable of being that God supreme residing outside of the created world.

In a preceding section in which some contemporary context to the study of chemistry was given, I remarked how a traditional natural theology cannot be applied to a physical science for which the objects of study are not fixed, not immutable, nor created 'naturally' in the sense of having arisen without human intervention. To some, Process thought could be applied here in that it is useful when framing a 'speculative philosophy' around ever-varying objects or objects whose ontologies are best expressed in terms of their necessary changeability. This I find an unconvincing argument: chemistry makes use of often hidden yet still very real fixed laws or rules of combination and transformation, to create new compounds and associated productive processes. Each such 'product' then has application and utility conferred upon it precisely on account of its structure and attributes: properties which are in that sense immutable in their application. There is no need to invoke Process thought to accommodate a newly discovered efficacious compound, metaphysically.

It might be useful to note in passing that whereas in this section I have been speaking of Process thought as it impacts theology, Ross Stein has made a case for describing chemistry in Process terms from a purely secular view point (Stein, 2004, pp. 5–22). Because Process thought sees itself as a systematic treatment of reality as already explained, to in a sense 'qualify' as a fully Process-driven analysis, a system must exhibit several traits: yes, Process thought sees the reality of a system as expressed in its dynamic properties as opposed to any static ontologies, yet it should in addition be shown to have an identity formed teleologically: what a something is, is what it is in its 'becoming'. Many chemical systems exhibit a profoundly dynamic quality where relationships between individual objects are constantly being made and broken, and then re-made. The water in a dark closed container at 5 Celsius and one atmosphere of pressure is just such a system where hydrogen bonds between oxygen and hydrogen nuclei are constantly in flux. Yet overall it is simply a glass of water. It is not in any Process sense, 'becoming' anything other than a glass of water. It is perfectly true that, as Process thought teaches, it would be insufficient to describe a glass of water as simply a collection of molecules of H2O and that it must be more fully understood as a dynamic series of interactions between such molecules: but it is not 'going' anywhere, it is not 'becoming' something other than a body of water; the full reality of that glass of water is fascinating but is more readily described as a series of repeating processes, repeating endlessly, rather than a process directed towards an end. This is a very restricted example, but the reader may also wish to explore so-called chemical Redox reactions where a manufacturing process does not fully move from starting materials to product and waste yet rather achieves an equilibrium where, dependent upon the physical context, more or less quantities of product co-exist with starting materials and waste: the reaction itself is constantly happening in both forward and backward directions (see Appendix B for a brief introduction). Again, although it is undoubtedly dynamic, it is not moving decisively in any one direction yet rather constantly cycling about an equilibrium point. Slight variations in temperature, pressure, the medium within which the reaction is taking place and the concentrations of reactants, can all affect the relative amounts of starting materials, product and waste at the equilibrium state. Thus and again, the identity of such a chemical reaction does not lie in what it is becoming, it lies in where it has arrived once an equilibrium state in the reaction is reached. The sense of wonder experienced by the human observer perceiving such a Redox reaction is in the physical arrangement of the laws of the natural world, which enable such mechanisms to be a reality in the first place. Yet again, in defence of Process theory, I believe it would be wrong *chemically* to describe a glass full of water as an object, or a collection of objects; its 'being' is indeed dynamic and any comprehensive *chemical* account of its reality could only be made in descriptions of the forever processes that are taking place in the glass. Thus I would not wish to be described as being 'obsessed with describing reality as an assembly of static individuals whose dynamic features are either taken to be mere appearances or ontologically secondary and derivative' (Seibt, 2013, p. 1), whilst cautioning that it is more challenging to describe the glass of water as 'becoming' something.

In considering the life and work of Christ within Whitehead's *magnum opus* it is I believe instructive to note that the text-string 'christ' occurs five times and only once as 'Christ' ,whereas 'god' occurs very many times across its more than three hundred pages. Whitehead first presented his ideas as part of the Gifford lectures of 1928. There is no sense here of expecting to have a personal encounter with Christ contemporaneously known as a person, who like us is of the earth and yet who is also Divine. Significantly on this point, Ian Markham in countering a 'naturalistic world-perspective' sketches a 'different framework within which to view life':

> We share the insights offered by scientists on the mechanics of the story, but see these mechanics as a part of a greater whole. The cause, the heart, and the hope of the universe are goodness and love. This is what a theist means by God. God is a being that causes all things to be, who has the characteristics of personhood in that s/he can decide, feel, act, and is good. It is an optimistic view of the universe. The universe on this account is not ultimately inanimate, but personal. (Markham, 1998, p. 19)

Markham at this point in the text adds a footnote 'This account of God is heavily influenced by process theology'. It is a profoundly anthropomorphic view of the ontology of God where the word 'personhood', signifying a human quality, is being used to qualify the being of God. And this analysis is not only true of Whitehead but also of Charles Hartshorne, who arguably as Whitehead's expositor made his thought more accessible. Of Hartshorne's interpolation of Process theology it is said that it is simply another 'rationalist dogma' and that the anthropomorphism of such 'remains unconquered by philosophical abstraction, even if it is spread thinly throughout the system' and that his (Hartshorne's) theology is 'irretrievably anthropomorphic' (Gunton, 1978, p. 222). And yet when Gunton speaks in the same place of the 'utter lunacy' of taking such 'rationalist dogmas' seriously I think he has gone too far: in the life of Faith there is so much more, indeed there is an entire dimension beyond, the notion that the Universe is simply a container for mere objects that arrived here by accident. To crush a serious attempt to recognise that this dynamism informs much of the natural world is unfortunate. Perhaps at this point it would be useful to distinguish between Process thought or metaphysics and Process Theology: there are useful insights to be derived I would suggest from the observation that objects are at the micro-level not fixed; that much of our experience speaks to us of a world in flux. Of those who have developed such ideas into a theology, it is plain that these are not orthodox Christian as I have said, however this should not allow us to negate the import of that initial observation.

In speaking of Process theology, note that this view of God as expressed so far here, is not Trinitarian: nowhere is the utterly necessary rôle of the Incarnation even alluded to, nor the divine 'personhood' of Christ or the Spirit. Once Christ is recognised for who he is, questions of the degree to which God feels pain or whether he can be appealed to, or whether indeed God effects change in the physical world, can surely be addressed in the facts of his birth, life, death and resurrection.

Yet for Markham it appears that purposefulness in Creation equates to personhood, with perhaps a hint of panpsychism being accorded to a 'something' of, in or about, the Universe. Yet as elsewhere in writings on this movement, it can be difficult to discern precisely what is being spoken of: elsewhere Markham speaks of process being 'the means by which truth is discovered'. Further exploration reveals that Markham is speaking of the flourishing of community as Christians work-out their calling of service to a multicultural world. Christ is not mentioned: it is all somewhat hopeless (Markham, 1998, pp. 125–129).

There is no sense in which the chemist enquirer into the Christian faith, in dialogue with a Christian might get that sense Brümmer implied, of the 'penny dropping' and of having that personal encounter with a personal God who has an interest and involvement with them contemporaneously in their life. And so I conclude that forms of theology deriving from Process thought are not suitable candidate theologies when looking for chemistry to inform a natural theology.

3.2.7 Attempting to Resolve the Paradox of Simplicity Over and Within Complexity: how Should *Order* be Understood?

The natural world is essentially and fundamentally an entity of indescribable complexity, yet those who seek to probe its mysteries frequently resolve its intricacies in their given field through descriptions that revel in simplicity as a necessary and desirable virtue. Those who are opposed to theistic explanations of the origins of creation have, as is well known, at times preyed upon the apparent paradox of the God who gave birth to such overwhelming complexity being also understood in simplicity in Christ and His earthly ministry, as reported in the Gospel accounts. We have above proposed one solution to this conundrum in suggestion that the phenomenon of *pattern* might be understood as a form of reductionist grid over complexity, to forge the similitude of simplicity. Yet we can not speak of *pattern* within God in any meaningful sense (although as we shall see, we can speak of a pattern in manufacture). Our previous description works well for natural phenomena yet provides no answer when trying to understand how God in transcendence mediates a bridge between the faith world which through God's grace is permitted to know Godself, and the secular world which seeks for *proof* in some sense, of God's existence.

Thus whilst McGrath, as noted above, speaks of *order* in the natural world as indicative of the God who gave birth to it, any person who seeks to distinguish the man-made from the natural within any picture of the natural landscape, simply looks for straight lines and ill-fitting materials to distinguish that which humans have imposed, from that which has arisen of its own accord. Thus the *order* being spoken of is that of rule and mechanism and not one of physical expression, of shape. Interesting Polkinghorne places in apposition the contrasting views of cosmic history of 'unfolding creative improvisation' *versus* 'divinely preordained score', and in Christ I would suggest that *both* are in play (Polkinghorne, 2004, p. 54). Later Polkinghorne goes on to resolve this very point:

> The Father is the fundamental ground of creation's being, while the Word is the source of creation's deep order and the Spirit is ceaselessly at work within the contingencies of

open history. The fertile interplay of order and openness, operating at the edge of chaos, can be seen to reflect the activities of Word and Spirit .... (Polkinghorne, 2004, p. 81)

A natural theology arising from an anthropological or merely human appreciation of order (and so making use of the so-called 'Arguments from Design' – teleological arguments) is not wholly sufficient for an understanding of the place of chemistry within the natural world. Chemistry, as has been remarked above, makes things and delights in the making of them. It works with the natural order to manipulate that which is given in the natural world, the tools of reaction mechanisms and the raw materials, to create new materials for the use of humanity, as a sign of that 'ontological openness to the future' spoken of by Polkinghorne and as quoted above. Brooke and Cantor (2000, p. 314), together with other commentators, have remarked how this means that a chemico-theology has not enjoyed a 'high-profile', yet such an approach arises from a misunderstanding of the role of human-created or ordered artefacts including medicaments, in the forging and fulfilling of what is apparently the will of God.

It is now necessary to digress briefly to demonstrate that the Christian scriptures do in fact sanction humanly orchestrated artefacts and efforts in the gifting of God's blessings. The point of this digression is to show that it is wrong to disqualify chemistry solely on the basis that its endeavours rest on that which humans have made:

- Towards the end of the Book of the prophet Ezekiel, where a lengthy vision of a restored Israel is set out, and the glory of the Lord which had previously been shown to be departing the Temple has returned, a river is described lined with many trees, the leaves of which are to be used 'for healing' (Ezekiel 47.12). This account is echoed in the final book of the Bible, Revelation (22.1–2). Thus in a time of perfection, it will still be necessary for people to be preparing plant extracts for medicines.

- In the well-known account of a king's healing, a poultice of figs was prepared (Isaiah 38.12)

- If the preparation of food might be considered a form of chemistry, it should be noted that Christ ordered a girl just raised from the dead, to be given something to eat (Luke 8.55).

- And finally and perhaps most importantly, humans in Genesis 1 are told to 'rule over the earth and subdue it' (Genesis 1.26). This instance as well as the command to build the desert tabernacle according to a God-given pattern (see Exodus 25.9, 10) including the instructions for preparing for example the incense, would seem to indicate that followers of God are expected to build and create new materials, based upon patterns which God has already indicated. (see note to Genesis 1.26 in the NET bible)

It is God's insistence on people manufacturing according to a 'pattern' gifted by God which frees such efforts of humans to work with the materials given in the natural world, from accusations that this line of reasoning is merely Process thought by another means. The following of a 'pattern', indeed a God-given pattern in manufacture, would seem to be more closely allied to the saying of Jesus that 'without me you can do nothing' (John 15.5).

In contrast, an insistence on a natural theology being reflective of the order God placed upon the chaos He faced prior to Creation, speaks in favour of predictability within the natural world. Yet when faced with the complexity of that world, prediction becomes immensely difficult, something we suffer from every day in the coming of for example the weather, or earthquakes. Within chemistry, outcomes in experiments are often so sensitive to fluctuations in temperature, pressure, solvents and concentrations, that reliable predictions of outcomes are again immensely hard to make. There is great complexity, which we seek to rule over through the formulation of models, which in effect *create patterns within* complexity to give a degree of predictability. The science of chemistry thus attempts to inject a semblance of order or intelligibility through the application of rigorous conditions upon experimentation, thus controlling outcomes to a degree. And 'even when we understand it we are still in wonderment' (Lazslo, 2003, p.12). This 'wonderment' then is a sense of the intense intellectual stimulation when confronted not by one's own ingenuity, but by the intricacies of the natural systems which together uphold and inform each specific set of experimental conditions and then give rise to outcomes; perhaps also there is wonderment at what it takes to control experimental conditions to achieve certain outcomes: here the complex has been 'simplified', until the next challenge approaches. And now suddenly within this 'wonderment' those experiencing it do in-part make use of that 'bridge' between these two worlds of understanding we spoke of at the beginning of this section, and in discussions of an aesthetic response before that.

3.3 Beauty as Bridge

Here I state the reasons for utilising intimations of beauty as a bridge between chemistry and a natural theology.

As has already been said, such a theology must be described in language that is intelligible to those who do not yet know Christ by faith, such that it may reasonably be seen as a destination of their enquiries into any natural theology informed by chemistry. Furthermore such a theology must in addition be perceived as lying fully within the orthodox Chalcedonian confession of the majority of Christian traditions. Yet such a theology can only be seen in the distance: it might be held by the one putting forward natural theological arguments and yet following its arguments does not constitute – as we have said above – the sum total of a valid Christian faith for the non-Christian honest enquirer. The latter type of Faith is the gift of God and only in His gift: it is not in the gift of human reason.

Pannenberg in explaining a particular issue in an introduction to systematics, makes use of a phenomenon within the natural sciences to propose a theological understanding of God's actions in the natural world (Pannenberg, 1991, pp. 46,47). The point here being that an element of a rational understanding of God – a theology or component of a (systematic) theology – is being reflected back onto a well-known phenomenon in the physical world. This is important because it suggests that theologians with some understanding of scientific methodologies might bring into play aspects of an understanding of the Christian God in order to validate a comprehension of phenomena within the natural world. Again it must be stressed that we are explicitly not suggesting that this amounts to mutual agreement – that scientists would see the proof they require and that those that are not of the Faith feel compelled to become Christians on that basis – merely that a mechanism for dialogue in explaining similar phenomena has been provided between the disciplines of science and theology. (It will be remembered that chemists have within their own discipline proposed that *similarity* be used in the same way: where research chemists agree that where two processes are similar they allow for similar predictions or inferences of outcomes to be made).

Thus it is rational to assert that, that which Christians theologians and chemical scientists – speaking collectively – both claim to be an appreciation of the aesthetic importance of a given entity within their mutual fields of study, – this ability – can be understood as a glimpse of the (Christian) Transcendent; the language is the language of beauty which as we have shown is mutually intelligible (even when they may not agree over what constitutes 'beauty').

3.4 An Expanded Vision of Beauty

3.4.1 Beauty in Art

This section expands upon the understanding of beauty in order that when a definition is proposed below I can be confident that the given definition encompasses a view of beauty that is not too narrow.

Whilst art is anecdotally taken to mean classically such artefacts as paintings, sculpture or buildings, or performance in music and dance and now latterly might include the moving image as well as performance, I have as yet said nothing of the 'artfulness' of constructions in for example software or forms in plastic materials and even representations existing purely as fixed digital images.

Consider the human response to for example a classical Greek sculpture of the naked human form, perfectly executed. Consider the case where the original in its life-size rendition is called 'beautiful' by an art critic and then the photographic image of that sculpture is seen by millions, many of whom pass comment for example on-line, that they too consider it beautiful. Makers of touristic merchandise now reproduce that same form in miniature, many thousands of times. In all likelihood no-one would consider these reproductions to be beautiful. Would a life-size replica in plastic or fibreglass be considered beautiful? Probably not or at least there would be a question hanging over it. From this example it can be seen that beauty has to do with not only perceptions of shape, proportion, skill of execution but also novelty – a sense of uniqueness. One Rolls-Royce motorcar might be considered beautiful but forty in a row might not [and Edith Wyschogrod (2003, p. 82) addresses the tensions in this issue]. Consider a painting. Sitting in front of it for the first time, leaves the visitor mesmerised. Forced to sit in front of it for hours at a time, their attention might begin to wander. Yet if that same visitor chose to gaze on the object several times a week over a period of time, their engagement with it is renewed on each occasion. Beauty has to do with shape, proportion, skill of execution, novelty and a fluid *constant re-engagement* by the watcher over whatever it is that is being observed. The person in life, in conversation, in communication, is considered beautiful this morning, this afternoon, tomorrow, but their static image will tire after a while. An appreciation of beauty is not static, it is a *precession* of the conscious mind as it engages with the beautiful; it is a moment-by-moment apprehension and then re-apprehension of the beautiful, almost as though the moment is being renewed constantly:

> "The judgement of beauty, it emerges, is not merely a statement of preference. It demands an act of attention. And it may be expressed in many different ways." (Scruton, 2009, p. 15)

and it is both the sense that there is a real something 'out there' as well as a whole variety of ways in which that existent beauty is expressed, that I wish to emphasise.

Beauty is therefore about a something that 'reaches to the underlying truth of a human experience' – it is ontologically real as part of the full or complete process that probes the entirety of a human event: it is indeed 'rationally founded' (Scruton, 2009, pp. 129, 197). It is noticeable when it is absent and whilst it may be exhibited by the apparently statically beautiful (for example a painting or a sculpture) it is in fact part of the constantly living and probing nature of the human condition, and as such always in motion. For this reason and as I have already said, because the static image of the beautiful may tire after some time, yet that any re-engagement with the beautiful, can similarly re-awaken the sense that one is engaging with the beautiful, apprehensions of beauty are necessarily dynamic. There is the observer, the one pronouncing a judgement that something is beautiful, there is the 'something' being declared beautiful, but there is also the sense, the explanans, the living moment and typology, that holds that consideration in being. This must be so, else we should expect there to be one day a chemical injection, drug or treatment, that the moment it is effective would see all things perceived, as being beautiful. Anecdotally I assert that this is most unlikely ever to happen: the movement of a footballer, solid cultured objects and forms, mathematical equations, song, images, memories of conversations, the shape of a smile, the list is very long and all have at times been considered beautiful. It is not the entity under consideration itself but the manner in which it is perceived that gives rise to the evaluation that something is beautiful. The observer is in a veritable *conversation* with the observed, and both are affected by the relationship. The one is now called 'beautiful' and the other is transported in his/her appreciation of that beauty. At times this may give rise to pleasure, or to a sense of wonder, or to awe bordering on fear or again other responses, emotions even, but as Graham Ward remarks, the beauty of it is a 'thing' of itself which then gives rise to a response: 'beauty is an operation, a co-operation; it is not a property but the animator of the properties of an object' (Ward, 2003, p. 38).

An object declared to be 'art' is subject to beauty as the 'animator' of its properties, as Ward tells us above. Art therefore is a revealer of beauty, although by no means the only such:

> Art .... stands on the threshold of the transcendental. It points beyond this world of accidental and disconnected things to another realm, in which human life is endowed with an emotional logic that makes suffering noble and love worthwhile. Nobody who is alert to beauty, therefore, is without the concept of redemption – of a final transcendence of mortal disorder into a 'kingdom of ends'. In an age of declining faith art bears enduring witness to the spiritual hunger and immortal longing of our species. (Scruton, 2009, p. 188)

In showing that chemistry may be beautiful or more accurately a conveyor or portrayer of that which is beautiful, it plainly does also enable someone who is 'alert to beauty' to stand on the 'threshold of the transcendental'. Derkse makes the same point as I note in chapter 2 above. In addition Scruton here makes the connection between art as something *static* (although its appreciation as I have said is not) and the *motion* of all humans towards the eternal. It is therefore reasonable on this basis alone to expect perceptions of beauty to lead people towards the thought that there may be a God.

3.4.2 Beauty as 'Other'

It is now necessary to expand my treatment of beauty by discussing beauty as transcendent or 'other'. This is so because as we have seen in the discussions on chemistry and as we will see again in a review of contemporary chemistry research papers below, when chemists speak of beauty it is frequently not about something which can be seen but about something that is interpolated through various measuring or modelling techniques.

If beauty were like other things we know, we experience, we can manufacture, it would not be beauty. Beauty is gift as well as mystery; it is if not wholly unique then it is uniquely given at each place and moment where beauty is received as gift. As Edward Farley implies, it is 'other':

> a distorted egocentrism would use beauty to allay the anxieties that come with self-transcending temporality precisely because beauty in the reality of its atemporality points away from the temporal self to an 'other'. (Farley, 2001, p. 92)

Farley also leads us away from the aesthetic way of life which in fact subverts Christ's beauty and denies that beauty has to do with things which are simply pretty:

> if beauty is there for one's use, if it is something to be 'experienced', then it must simply give pleasure. And to be oriented only to beauty's pleasure is to suppress the pathos that comes with the chaos, destabilisation and the mystery of beauty. These elements are present both in the classical view of beauty as harmony and in the Edwardian view of beauty as benevolence. To make 'pleasurable experience' beauty's point suppresses ….. beauty and, like Philistinism, reduces beauty to the pretty….
>
> To make beauty everything, to press for its gifts, is to lose beauty …
>
> … aestheticism is a child of the distorted imago
>
> [To replace beauty with aestheticism means] the abandonment of beauty as an ordinary and relative good for beauty as something able to be secured and found. (Farley, 2001, p. 92)

This is most important for if beauty were something that was only accessible by the privileged few (for example those leading an aesthetic way of life), or perhaps only by those who had access to some private *gnosis* or again only by those of a particular degree of taste, it is difficult to understand how the perception of something as beautiful could have Christ, who came to call humans from every 'tribe, language, people and nation' (Revelation 7.9) as its source.

Having established that beauty speaks of the 'other' and is accessible to all, James Fodor shows that it can disclose new ways of seeing:

> In both its verbal and nonverbal forms, great art discloses a world that is startlingly new and strange—at least in contrast to the world most humans are accustomed to inhabiting. Many of us pass our days in a state of enchantment, absorbed in a world of illusion, fear, evasion and self-deception. Art interrupts and unsettles these 'normal' patterns of seeing. (Fodor, 2008, p. 193)

From this it can be seen how important an 'agent' such as chemistry is, as a vehicle for displaying beauty, since the place of its finding is so often also 'chaotic and destabilising', and certainly frequently 'startling new and strange'. As for this strangeness, some of the papers discussed in Chapter 4 demonstrate a clear degree of 'underdeterminedness'. The degrees of uncertainty often associated with discerning precisely what it is that one has found in chemical research, lend themselves to such a conclusion, as does the air of mystery surrounding where one's current position will allow one to go next.

Thus from an entirely different manner of approach we have similarly been able to establish that to discern beauty within one's journey, in whatever part of one's journey, is to have encountered the transcendent, the religious and the ethical as a result and to have touched God in some way as that 'Other':

> The Spirit continually draws us forth in acts interpreting the mystery of that which, in being itself, is also an intimation of the mystery of otherness. That at the heart of the beauty of God lies a profound irony, that irony might be understood as an intimation of divinity, leaves each of us suspended in so much that is understood (re-cognized) while not being grasped. Suspended in the experience of seeing through a glass darkly, laboring in an ineradicable hope, and glimpsing in the beauty of God that there is nothing, then, that cannot be redeemed. (Ward, 2003, p. 65)

This remark then neatly not only describes the *telos* of our *conversation* with our non-Christian friends and colleagues but also validates our hope that God might graciously reveal Himself to them in the beauty that they have in hope, encountered in their investigations; a beauty that is a signifier of Himself; a beauty that is indeed true and at its root similar, whether it is encountered in fine art, nature, chemistry or apprehensions of God in Scripture.

3.5 Conclusion

A survey of chemistry has been provided including aspects of its history, an understanding of some of its methodologies and an indication of where the discipline touches upon metaphysics. In addition, part of the objective of this chapter was to mark-out something of the contrasts between the natural science of chemistry as opposed to those of mathematics and physics. It would be rare to speak of for example 'theoretical mathematics' since the practice of research in mathematics is frequently entirely theoretical. Yet 'theoretical chemistry' is an important and flourishing branch of the discipline. For all that and in direct contrast to mathematics and to a lesser extent of physics, chemical research has largely to do with the physical act of experimentation, with the practice throughout one's time as a researcher, of using the techniques and expertise acquired in a time of apprenticeship. So much of chemistry relies on the physical 'doing' of it. Adequate explanations often rely on a dynamic synthesis of *a priori* considerations usually developed by theoretical chemists, together with models, similarities, experience and inference practiced by those who engage in the physical work. The results of the research frequently elicit a sense of wonder and of beauty. This latter is arrived at not merely because of that which can be apprehended by the physical senses, usually sight and smell, although in the case of plastics and modern fabrics through touch as well, but often is appreciated internally as a response to compounds and processes which can not be seen but only inferred.

A theological accommodation has been established which is suited to a view of the natural world as *worked-with* in the chemical sciences. Such a view is also shown to be compatible with living in the contemporary everyday: a living that allows us to work through uncertainty and disappointment as well as with wonder and beauty, fully cognisant of the 'rock from which we are hewn' (Isaiah 51.1, Romans 9.30–33). We have seen how the language of aesthetics allows for mutual intelligibility between scientists and theologians in that it is recognised that in speaking of perceptions of beauty, they are speaking of the same thing. In order to facilitate a fuller exposition of beauty leading to a definition in chapter 6 below, I have also expanded our present vision or understanding of beauty through a brief survey of its apprehension in other disciplines.

Foster's and other's work as explained in chapter 2 have allowed us to accurately peg our natural theology as being not deist but specifically Christian, through a revelation of the love of Christ mediated in the here-and-now through the Spirit, as being that which is visible in the wonders of creation (Proverbs 8.30). Such 'wonders of creation' enable the natural theologian, using a particular expression of the argument from design, to rationally make the case for the hand of the Divine, being visible in the embedded frameworks shaping the natural world. Thanks to Derkse's insights it is possible to recognise the simplifying insights of the human mind making sense of the undoubted underlying complexities seen in nature. Throughout we concur with the work of Plantinga as developed and then extended in chapter 1, about the importance of clear thinking in the discernment of warranted truth.

We are now ready to explore certain recent research papers in chemistry, with a view to showing how in practice certain of these features of the discipline find their expression in the reports of the research.

Chapter 4: A Selective Survey of Current Organic Chemistry Research

4.1 Purpose of the Survey

The role of the aesthetic in chemical research has been explored elsewhere in this thesis and as such there is no need to prove its influence. This book is seeking to explore the usefulness of chemistry in a natural theology and whilst it might arguably have been possible to outline such a contribution from published sources it is nonetheless of interest to explore how the current use of language in chemical research supports aesthetic considerations most notably estimations of beauty. Of course what is of particular interest is that whereas perceptions of beauty in for example the graphic arts, architecture or music are immediately available to the senses, those in chemistry are not. In the latter case the entity or the process which is being assessed is largely or often invisible, with perhaps colour or smell betraying something, but of course nothing (except perhaps hints) of structure. Other characteristics of chemical research have been described and it would also be useful as part of this present survey to illustrate these with examples from current research. Such characteristics include the use of many and varied sampling and analysis techniques to provide a full description of the chemical event or structure in contrast with other scientific disciplines notably mathematics or physics where the descriptions are more usually in terms of equations. The practise of chemical research has been shown to be akin to that of craft as much as science and this remains true as the examples show. Certain of the research papers sampled here involve the creation of entire sets of related or similar compounds where analogous target structures or other related characteristics gave the researcher the impetus to attempt analogous processes to synthesise them. This latter illustrates the use of similarity in chemical research as being truth conducive.

4.2 The Selection Criteria

By way of explanation for the proposed selection criteria, please let the reader note that the author's MSc was in the area of reaction rates in organic chemical reactions. This is an area of research known as Physical Organic Chemistry. The MSc course itself also involved some presentations in the area of organic chemical synthesis. The particular reactions being studied were performed using relatively simple compounds in stand-alone reaction vessels i.e. they were not part of any biological system. The purpose of the MSc research was to test the boundaries of certain rules that had previously been put forward as governing such reactions.

In order to ensure that any research papers utilised in this current book are easily accessible, they have been chosen from the Royal Society of Chemistry online resources. This will ensure that the published research is peer-reviewed, authoritative, with the subject matter being non-trivial and in that sense reflective of current research concerns.

It would seem sensible to include papers from a variety of nationalities and research centres, as far as this can be determined i.e. not all from a single research group or detailed area of research.

Summary of Criteria

Taken together the criteria for selecting recent research papers in chemistry to be included in this volume are:

- Publish date not before 2013
- published and available online by www.rsc.org
- no more than 15 papers
- in the area of organic chemistry research and as far as possible concerning the non-biological synthesis of new compounds and/or involving a physical chemical component
- involving a variety of researchers
- not requiring any other selection criteria other than those outlined here.

4.3 The Papers

A set of contemporary chemistry research papers will here be referenced and discussed.

1. Koch, et al. (2015):

It would appear to be a report of the highly successful use of a novel new catalyst delivered in the very dry language of such chemical research. The use of such a catalyst delivers both environmental and cost benefits. Further examination illustrates the excitement felt by the team with their work, as well as their disappointments and surprises. It must not be forgotten that all such work is carried out on 'the shoulders of giants' in that it is almost always the development of previous work, techniques and practices and in this particular case this fact is demonstrated by references to those who have gone before as well as earlier research papers. One might validly ask what there is that might be unexpected or surprising about such work given the huge amount that has gone before. Yet here there is both surprise and failure. The physical arrangement of the diagrams and the ways that the graphics are arranged on the page illustrates the use of pattern, symmetry and dissonance, in the pictorial illustration of results.

2. Cordonnier, et al. (2014):

This research reports on a novel solution to a particular problem in this branch of synthetic organic chemistry. Within a multistep creation process certain types of product traditionally have required sub-components of the product to be created very early in the stepwise process. This latter is necessary because these particular sub-components require highly reactive conditions for their manufacture and should these methods be used later in the process, work done to add more sensitive sub-components later in the process, could easily be undone. The target or product compounds are not being created because of an immediate need. This present work is pure research where similar target compounds already exist and are well-known biologically active compounds. Thus it makes sense to attempt to develop novel methods of synthesis avoiding the traditional pathways which are unduly restrictive. There is within the writing of the paper, a readiness to appreciate earlier experimenters' work in this area, variously describing it as being 'elegant' and 'impressive'. Steps within the overall process are described at times as going 'smoothly' or 'uneventfully', both terms anecdotally one might not expect to find used to describe simple chemical processes, which after all may be perceived or inferred but not seen.

The chemical reaction mechanisms at work in the individual steps of the multistep process are at times described as though they are known accurately and at other times the phrases 'likely due to', 'possibly due to', and 'potentially due to' are used, indicating that there is a degree of confidence but insufficient proof.

The collaborators are not slow to probe and investigate further when for example their reactions yield unexpected results. They engage in this exploration I would suggest, irrespective of the success of their main work, perhaps demonstrating that this work is a voyage of discovery. At one point the accurate prediction of a likely result caused the researchers to speak of their 'delight'. In a final remark they speak of the 'rich reactivity' found within their particular specialism.

The paper is noteworthy for its lack of use of complex quantum mathematical modelling tools in the elucidation of mechanistic pathways, a common feature in several of the papers being considered here.

This paper, though admittedly complex, provides ample illustration of the aesthetic appreciations at work in current chemical research.

3. Hancock, Kavanaghab, and Schiesser (2014):

This paper is noteworthy for its particularly rich use of aesthetically-charged language. The writer almost immediately express 'astonishment' at the growth in their particular specialism; the level of activity here is described as displaying 'waves of intense activity'; pitfalls are described as 'demons' and successes as leading to a period of 'prosperity' in which the specialism 'blossomed'. In common with papers elsewhere, work by another researcher is praised for its 'elegance' and established common processes are described as previously finding a 'comfortable home' in the 'toolbox[s]' of such workers, leading to the creation of 'shiny new toys'.

This paper also demonstrates a sustained attempt to relate current experimentally derived data (meaning completely *a posteriori*) with results obtained from previously constructed computational models, to validate the proposed chemical process mechanisms. Such models have in the past been created out of other experimental work allowing an inference to be made about the possible mechanism of a new process, where the model and present experiment are in close agreement. This latter is illustrative of the tendency noted above in current chemical research to allow for similar results across experiments and experimenters, to be used to infer, a similar reaction mechanism by inference.

4. Kaufmann, et al. (2014):

This paper details the creation of a sequence of, as the researchers admit, 'fascinating' compounds whose overall shapes may be likened to a tripodal needle and thread. Each of the three 'needles', which are actually joined at the top of their shafts to form a single compound, threads through a separate circular structure on another single compound. These compounds are composed at the extreme micro level out of various chemical subcomponents. It is little wonder that the researchers report that all of this is 'quite remarkable' which of course it is. Their language betrays the wonder at their achievement. The reactions themselves are quite lengthy: one is reported to have taken over two weeks. In contrast some of the other reactions described in the papers being surveyed were taking place in timescales less than a thousandth of a second.

As might be expected anecdotally, the two parts of such compounds, the pin-compound and the receptor compound can be made to move closer together and further apart, by the introduction and removal of further reactants. Such motion, as outlined by the researchers, could form the basis for a molecular engine. The degree to which all of the reactants move in a concerted manner at the introduction of these further reactants is described by the researchers as being 'highly interesting to assess'.

In addition to the considerable and noteworthy success of the synthesis itself, the diagrammatic presentation of the report is very fine.

5. Li, Huanga and Wang, (2014):

This paper whilst it continues to a degree the theme established in the other reviews of the researchers expressing pleasure and delight at their success, also addresses another aspect of current chemical research: the way in which it demonstrates the quoted maxim of 'chemistry being what chemists do'. There is no *a priori* shaping of the pathways to successful syntheses of the target compounds and indeed at one point the workers state plainly that a given attempt failed to produce any product compound *at all*. The paper does demonstrate an unrelenting drive to achieve success in synthesising a range of similar compounds with various approaches being attempted before the desired results appeared. This is very much empirical science very starkly drawn. The researchers show themselves to be true artists in their field with a fine sense of what elegance in method means. This they achieve themselves.

6. Berger, et al. (2015):

This paper is included here as a perhaps more extreme example of how multiple models or methods of description and analyses are deployed to give a more complete picture of either a single process or series of product molecules. In this paper many such techniques are used including nuclear magnetic resonance, x-ray crystallography, mass spectroscopy, ultra-violet absorption spectral data and fluorescence spectra. Further, the paper also contains graphics representing the proposed molecular orbital shapes at certain points in the products. In this way chemistry is shown to be quite different in character to either mathematics or physics, and in-spite of the sophistication of the techniques employed a single reductionist description of the entities involved seems ever further away.

In contrast to some of the other papers reviewed here, this group of researchers is very well resourced and has access to many analysis techniques. This illustrates the point made elsewhere that the breadth of current chemical research is very wide.

7. Tuna, Sobolewskib and Domckea, (2014):

It has elsewhere been alleged that chemistry can not contribute towards a natural theology because the route to a natural theology is via metaphysics. Since, as alleged, there is no metaphysics associated with chemistry *ergo*, there can be no natural theology associated with it (Fraser Watts 2014, private communication). This present paper illustrates the challenge presented by such assertions in that on two occasions statements with clear metaphysical implications are made (see the relevant quotations given below), and no comment is passed. Thus whereas I have elsewhere amply demonstrated the metaphysical content to be found in chemistry, if the commentator chooses to ignore it, the lack of theological insight is not surprising. The two instances in this paper are as follows and relate as the authors say to the 'molecules of life':

> The fact that the spectra of the energetically most stable conformers or tautomers are often not observed is a clear indication that ultrafast (sub-picosecond) radiationless excited state deactivation processes prevail. It has been argued that these ultrafast excited-state quenching processes provide biological matter with a particularly high degree of UV photostability (p. 39)

and:

> The quenching of deleterious photochemical reactions by ultrafast internal conversion is believed to be decisive for a high intrinsic photostability of DNA bases as well as amino acids and peptides with aromatic chromophores (p. 39)

Both of these statements might reasonably be expected to elicit a passing reference as to the process by which such extraordinary provision to protect living matter from UV damage, came to be present, yet such comment is absent. It might be argued that this research is not concerned with such matters however such a conclusion is brought into question in the calling of glucose, being the molecule being studied, one of the 'molecules of life' elsewhere in the paper.

8. Lang and Smith, (2014):

This paper is concerned with the accurate modelling of distances between the atoms of individual elements involved in compounds and solid structures. This 'bonding' is important because as the paper points out:

> Covalent and ionic radii are used in structural chemistry and molecular modelling. Reliable data of ionic or covalent radii (or internuclear distances) can serve as a rough guide to the magnitude of steric effects, how reactions may occur and on the stability of compounds. (p. 3355)

There is very little aesthetic language in this paper however it is noteworthy in this present study because of:
 – The combined use of empirical, estimated and theoretical (*a priori* quantum mechanically derived) data, to create substantial tables of bond lengths;

– The demonstration in a specific instance of how an aesthetically pleasing result was eventually shown to be inaccurate:

> The "free electron/electron sea" model, although very elegant and convincing at the time before many sophisticated experiments were made possible to test it and before quantum mechanics was firmly established, is shown to be inadequate. We consider that "band theory" provides a correct "theoretical description" of metallic structure. (p. 3367)

This is an example spoken of elsewhere how beauty need not necessarily give rise to truth. More generally it is noteworthy how this information about bond lengths is in almost universal use across huge areas of chemical research and yet as the authors note:

> There is a proliferation of different series/sets of ionic and covalent radii in the open literature. Goldschmidt and Pauling used different methods to estimate ionic radii in the early part of the twentieth century. Amongst the various sets of ionic radii, one produced by Waddington is fairly commonly quoted and a widely known set is put together by Shannon and Prewitt and later on improved by Shannon. However, it was pointed out that there are some impressive discrepancies between these sets and they do not fare well on certain statistical tests. Observed radii differ substantially from the commonly known sets of radii with a few exceptions. (p. 3355)

From this we understand that in spite of the fundamental importance of this information, the fact of much of it being inaccurate, has little bearing on the vibrancy of the industry. Once again the importance of the empirical nature of much chemical research, of it having a 'trial and error' approach, is shown. Chemists are engaged in an exploration of the possible, continually adding to human knowledge and achievement simply by applying the hard-won skills of artisans and crafts people in the manipulation of techniques and starting materials. Having created new compounds or established new techniques, these are then poured into or applied to, new explorations. As a result there is a constant unfolding of, a revealing of, what nature can produce if the requisite conditions are met – and only if these conditions, these rules, are met. It is this that invalidates the fears of Brooke and Cantor (2000, pp. 315, 319, 329) that chemists might be seen as creating 'unnatural' things, by themselves and without recourse to the Creator: yet researchers can only create what Creation and its Creator will allow to be created.

9. Nguyen et al. (2015):

The possibility of creating a new compound with a novel structure that has nonetheless no immediate uses is often sufficient cause in the chemical sciences to pursue the goal.

Imagine a football consisting as it does, of panels of six and five sided flat shapes. At each 'corner' or junction there is a carbon atom:

Figure 3: Fullerene

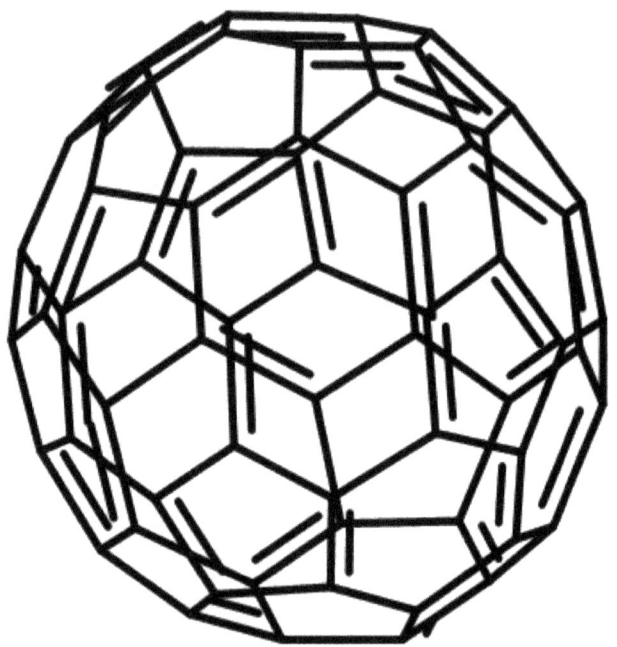

(See Appendix A for an introduction to the typographical representation of bonds). The diagram here is a cage of 60 carbon atoms and such 'Fullerenes' are known to exist in various sizes. It is obviously desirable to attempt to create similar compounds using other elements instead of carbon. Because of the unique properties of carbon certain of these fullerenes are remarkably stable compounds existing simply as black dust under normal conditions. Such stability is not found for all such fullerenes and this present paper discusses the use of certain metal atoms to 'stabilize and modify the structure and properties of the resulting doped cluster' made in this case out of the element boron. The use of the word 'cluster' indicates that not only cages are being created but also tubes and sheets. It is with such tubes or sheets that the utility of these compounds increases. The paper discusses the 'extremely rich structural features' of these compounds and states that the workers engaged in an 'intensive search' for new compounds, thus expressing some of the excitement associated with these explorations. This paper also amply demonstrates an aesthetic appreciation of the compounds themselves.

As an aside, of the original C60 carbon 'version', the original Buckminsterfullerene, Peter Walhout has to say:

> With a backdrop of aesthetic theory now in place, let us explore a specific example of sublime beauty in science. Richard Smalley, together with Harold Kroto and Robert Curl, received the 1996 Nobel Prize in Chemistry for discovering C60 (buckminsterfullerene, or "bucky balls" as they are affectionately known). C60 is a new allotrope of carbon in which sixty carbon atoms bond together to form a molecule that looks exactly like a soccer ball, with single carbon atoms making the vertices of the five- and six-sided panels that make up the surface of a soccer ball. Shortly thereafter a similarly shaped C70 appeared along with many other types of this new class of molecules now known simply as fullerenes. Included in this class are the famous carbon nanotubes, which can be thought of as graphite rolled into a narrow tube. All of these are undeniably beautiful, owing to their symmetry and unique shape. (Walhout, 2009, p. 769)

A Nobel Prize having been awarded for this discovery, surely none in the chemical sciences can be ignorant of the beautiful within chemistry?

10. Yu, et al. (2015):

This paper presents an extreme proliferation of compound names in their modern prolix forms. Yet at its most simple it represents a highly effective and novel methodology for the creation of a particular class of biologically useful materials, using Gold as a catalyst. The materials in question are relatively small compounds and are frequently used as precursors in further reactions. The researchers speak of being 'inspired' by their previous results to 'fine tune' a particular aspect of their process, which 'to [their] delight', yielded the reported results. As is not uncommon and indeed as reported also in this survey of papers, these researchers then set-about modelling further experiments to determine if their novel methodology could be more widely employed, not only in the specifics of their initial reaction, but also more generically across this class of compounds. In this context 'more widely employed' means being utilised for the addition of a greater variety of groups to the initial reactants. A table is reproduced in the paper to show all the various groups tested. Not all were successful. Where the new reaction conditions were not successful, the outcome plainly was not predicted and in fact consisted of somewhat of a mixture of products. The researchers attempted to determine the key process pathway and conclude that the reaction is 'presumed' to proceed in the way they state. The reasoning employed by these researchers which motivated them to proceed in the manner they did in their 'fine tuning' is not outlined, however given that it uses an established methodology (reported as 'Ellman's tertbutylsulfinimine chemistry') it is rational to assume that it came about as a result of an appropriate 'apprenticeship' in chemical methodology. This apprenticeship would then have given to these researchers a set of practiced paths to tread when requiring certain outcomes.

This paper in summary demonstrates indeterminacy, the importance as explained elsewhere of the didactic process of 'learning the practice of' chemistry, of non-formal reasoning leading to a successful outcome, and finally a keen sense of the aesthetic not so much in this case in the product, but more in the 'beauty' of the process.

11. Ilichev, et al. (2015):

This paper shows apparently an almost complete absence of any aesthetic motivation and is reviewed by way of contrast. It describes the creation of large complexes consisting of organic components combining with very large and rare metal ions, whose overall structures are then characterised using in the main X-ray crystallography and luminescence. Nonetheless it is notable for the use of reported colours of the various compounds as one method of characterisation. In addition the diagrams of the compounds are presented in such a way as to emphasise their pleasing structures and mutual arrangement of sub-groups. This paper is notable for the presentation of various theoretical models of energy transfer between various parts of large complexes. These are declared to be able to predict various outcomes until changes in structures cause the models to fail and news ones to be proposed. It is these energy transfers which give rise to the observed colours and luminescence.

Perhaps this paper could be cited as an example of those being confronted by obvious beauty then refusing to acknowledge it.

12. Prati, et al. (2015):

This paper illustrates a form of inductive reasoning. These researchers had enabled a particular methodology to work before in one particular set of cases: might it work more generally? This paper also illustrates an attitude to experimental activity in chemistry perhaps in the past associated with alchemy: different processes were attempted simply for the sake of trying them, without any reporting of the reasoning. Once again this illustrates how chemistry is often very much about the 'doing' of it, empirically trying differing methodologies in pursuit of a result, rather than the carrying-out of a carefully planned, logical sequence of steps, previously predicted to work theoretically.

As for the detail, the researchers report the effectiveness on particular reactions of introducing dust-like particles of gold and ruthenium set into supporting matrices of variously carbon and a type of glue. These processes make use of what is known as the 'catalytic' properties of these metals, where it appears they enhance and at times enable reactions by modifying the chemical reactants immediately prior to a reaction. Thus these catalysts are not products or precursor compounds and in fact are unchanged in the overall process. It might be helpful to envisage them working rather like a crowbar in the opening of a box or door: the crowbar is itself unchanged but radically improves the possibility of some form of result. The catalyst appears to take hold of and then 'offer-up' the reactant to each other, increasing their latent reactivity.

In addition to simply trying out different options in an attempt to gain a successful outcome, the paper also illustrates that the precise structure of these matrices holding the metal dusts in place, was not known and had to be elucidated. This illustrates the principle of indeterminacy in both reaction pathway, structure of the catalysts and likelihood of outcomes.

13. Zurek and Wojciech, (2015):

This is a remarkable paper. Unlike several of the others surveyed here it is quite lengthy. It displays in places a metaphysical approach to its investigations, its scope is broad and it makes several sweeping statements and promises. In that sense it is visionary. It suggests an aesthetic sense on the part of the researchers who scrupulously avoid the use of any such language excepting to say that in some places, findings are 'exciting'.

Epistemically this paper is significant since it appears to champion prediction based on theoretical understandings. Why should it be 'exciting' that 'theoretical predictions' are 'guiding experiment' (p. 2917)? Surely results may be exciting, as we have seen in other papers above, whether predicted or not? Is the power afforded through the ability to predict outcomes, more to be desired than the power that comes from an ability to create and measure? The researchers here also make reference to an experimentally verified prediction of several years ago in a related field that has little bearing on their present work, showing that this ability to predict outcomes features strongly in their value-system.

Some explanation about energy minima might be useful in understanding this complex paper. If someone were asked to predict where a free-rolling ball-bearing would come to rest in a geography of a pair of hills separated by a deep valley, there would be no hesitation in declaring that it would be found at the lowest point of the valley. If the floor of the valley were to start to rise there would come a point where the ball bearing would roll off the now-raised valley and move past the former 'hill top'. Conceptually, mathematical modelling of chemical systems can be used to predict stabilities in systems that are under stress: it may for example be possible to predict that at certain temperatures and pressures, new forms of familiar materials could be discovered to exist, simply because the modelling predicts certain energy 'valleys' at those points. This paper uses such understandings in its modelling. However in-spite of the importance to these experimenters of the reductionists powers of mathematical predictability, experience is also acknowledged by one of the researchers, to be an important factor in directing experimental pathways (p. 2919).

14. Rong, et al. (2015):

Having now reviewed several papers and understood something of the beauty and elegance which may be seen in the workings of various chemical systems, I here review a paper where the reader may decide for themselves whether something wondrous is being described. Most will be familiar with the action of tuning a radio to search for differing programs: each is being transmitted on their own wavelength and each becomes accessible as the circuitry is 'tuned' to receive a particular carrier signal. Thus a single piece of apparatus is used to listen to a variety of transmissions. Plainly it would be advantageous to be able to create new chemical compounds in a similar fashion: using the same procedures but with 'tuneable' reactants to create different compounds (see Appendix B).

    The researchers have demonstrated the creation of such a set of reactants that are made active through irradiation with simple visible light. These processes take place in the presence of such reactants that act as catalysts, meaning that they are themselves unchanged and may be recovered and re-used. Although the researchers themselves do not make use of any overtly aesthetic language in this paper, the images are displayed in a particular way showing the reactants at their best: symmetry is well displayed.

    The paper describes a series or set of catalysts all belonging to the same overall structure, with a central metal atom of Iridium surrounded by complexes of other groups and atoms. It is as these surrounding complexes are subtly altered in their constitution that differing 'strengths-to-react', so-called 'redox potentials', are achieved. The result is that for example it is reasonable to assume that catalysts of this general type when introduced into a reaction could do more or less 'oxidising' on a given input compound, depending on how 'strong' the redox catalyst is. On p. 144 the researchers remark that novel uses for these new catalysts are currently being investigated. I would suggest that the entire design that this research paper proposes and in part demonstrates, is suffused with a certain elegance and beauty.

The general criticism, mentioned above, of chemistry in relation to a natural theology, that it is of little use as it 'makes things' is again called into question once chemical research is studied in this detail. It is plain that the researches are struggling or wrestling with, that which is entirely 'natural', in the sense of being a 'given' by nature. There is no sense in which for example these workers are creating new rules of nature; they are uncovering news ways in which the unseen laws that regulate the created world, may be manipulated for the stated aims.

15. Dattatraya, Balu and Raghavender (2015):

A casual survey of published research papers will demonstrate the enormous quantity of work being performed by chemistry departments in the Peoples Republic of China. In attempting to show that the aesthetics of chemistry are not limited to one region of the world's research laboratories I present this paper submitted by a chemistry department of an Indian University. Many research departments or groups will specialise in a particular area. These researchers are no exception and have a particular interest in establishing routes to the synthesis of so-called indenes for which the base compound consists of a six and a five member carbon ring fused together. This series of compounds are known in the wild and the synthesis of one such andirolactone has been achieved several times in recent years yet not without some complex procedures. In this paper the researchers report how they have been enabled to manufacture this material via a new method, which they describe as being 'concise' (p. 161) and in very good yields. The compound in question exists in a pair of stereoisomers, these being compounds with the same atoms connected to each other in the same manner but with different spacial arrangements (see Appendix A). They used a catalyst which because of its toxicity was not thought desirable and are now investigating whether they can achieve the same results using the same methodology but with a better catalyst.

The researchers achieved their results through a process of exploiting a series of 'happy misfortunes' where a path that they had been expecting to work, produced unexpected results. They went on to exploit their new discovery and to their 'delight' discovered that it had the desired result not only in the one area they had hoped, but also more broadly (p. 160).

This paper therefore, together with the others reviewed here, demonstrates that the ability to appreciate and to experience delight in the workings of chemistry might be expected to be a human quality more broadly and not be culturally mediated. It is also pleasing to note that the intensity of effort required to both become a practiced researchers as well as an innovative exploiter of whatever 'nature' lays before the investigator, is also a trait that is found widely across the discipline.

A diagrammatic representation of the base compound indene:

Figure 4: Indene

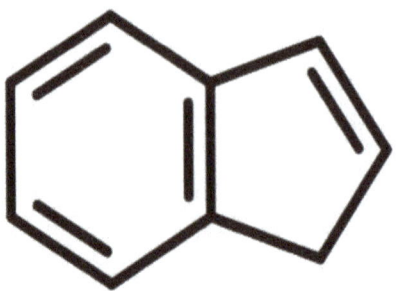

In this diagram, each 'corner' represents a carbon atom.

A 3D representation of andirolactone:

Figure 5: Andirolactone

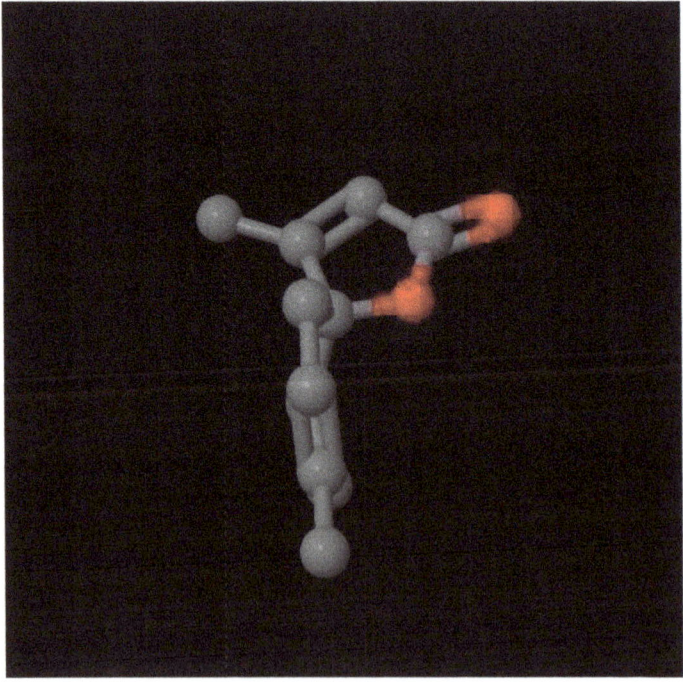

Where the two items in the top right of the diagram in red are oxygen atoms and those in lighter grey, carbon. The reader will note, and after reading Appendix A, that double lines in this representation are accompanied by a flat or planar structure at that point in the diagram. Single lines are associated with more angular dispositions.

4.4 Discussion

The objective of this small survey of contemporary papers in chemical research has been illustrative, so as to provide examples of what has been mentioned elsewhere in the text and particularly in chapter 3. The excitement felt by researchers in uncovering new process, compounds and pathways is plain in most of the papers surveyed perhaps exemplified by that of Hancock, Kavanaghab, and Schiesser. Yet such excitement can be found amongst researchers in many fields of human endeavour not just chemistry. What characterises such expressions of success in chemistry in particular? I would suggest that it has to do with an uncovering of variously: aesthetic qualities to processes and products, an unfolding degree of orderliness and 'fixedness of process' and also that satisfaction of honing one's skills as an experienced researcher such that intimations of possibility coupled with the skills of an experienced practitioner, produce pleasing results. Other scientific disciplines yield results where there is something existing waiting to be discovered but in chemistry, novel processes designed by researchers give rise to novel outcomes which are nonetheless not random but conform to some pattern which might at the start be only dimly discerned.

There are intimations also of various value constructs. For example, Koch, et al. (2015) and the group of Dattatraya, Balu and Raghavender (2015) place value on such characteristics as low cost, low environmental impact, a more concise process in comparison to the existing. Kaufmann, et al. (2014) describe something which mimics on a very small scale – an engine – that which is currently only commonplace on the large scale, thus placing value on miniaturisation. Chemistry might in these instances been seen as perfecting or improving upon existing mechanisms in ways which lend themselves to being thought of as more elegant or aesthetically pleasing. And of course these discoveries of that which is pleasing in this way, are not one-offs but fill the lives of such researchers.

Such patterns spoken of in the first paragraph above together with an aesthetically pleasing mechanism elucidated after possibly years of training, can result in the discovery of an extended pattern such that whole new groups or classes of compounds or processes can be uncovered. The work of for example Yu, et al. (2015) and Nguyen et al. (2015) describe such research in uncovering a sequence of similar compounds. The latter group also illustrates the propensity for researchers at times to explore simply for the sake of the beauty of doing so and with no immediate use of the compounds discovered in sight.

The experienced researcher has a varied and ever-expanding 'toolbox' of individual tools available to variously plan their research, to sample and elucidate the mechanistic pathways taken by reactants and to discern what their results are. Such a host of tools gives rise to a collection of descriptions so that pathways in a reaction can be described in possibly several ways using for example diagrams on paper, energy levels in mathematical equations and the movement of particles between reactants. Smell, touch and colour of products at times continue to be used to characterise outcomes. Different proposed pathways are not always exactly known and can be 'believed' or 'presumed' to be being employed in a given reaction, for example. Such underdeterminedness enhances the sense of beauty at the discoveries being made and contributes to intimations of elegance and wonder at the processes uncovered.

Simply because something is to an observer beautiful or elegant or a source of wonder might not lead the researchers to acknowledge that. The work of Tuna, Sobolewskib and Domckea (2014) illustrates this. There is the implication that those who want to be excited about their work, that those looking to appreciate the beauty and wonder in what they have discovered will frequently find it, yet equally that those who refuse to perceive this, will not. This sense of having to be open to the Divine action in the natural world in order to see it, is something that will be explored in the next chapter 'On God and Beauty'. By referring to ancient authors I will show that such scepticism and 'closed-ness' to intimations of the Divine, is not new.

4.5 Conclusion

Whilst admittedly not a large survey, it has nonetheless been possible to demonstrate several key points within this reading:

- The use of similarity and elegance to successfully predict successful outcomes, this meaning that these were 'truth-conducive';
- The importance of a form of apprenticeship in the career of the chemical researcher; people who through practical experience have learned how to do things correctly, so as to be prepared for success;
- Chemists giving many examples of the use of aesthetic judgement in the appreciation of their work; the word 'beauty' is not used, but the way this

appreciation is reported suggests nothing less than an often profound appreciation of the beautiful;
- Examples of underdetermined-ness in process where how the researcher got there is of less importance than the fact that they did, within acceptable environmental considerations;
- Examples of the use of both modelling as well as multiple sampling techniques, to provide an overall, yet frequently still inexact, view of the process;
- Examples of the use of both empirical and non-empirical study, being brought to bear upon *the same issue*.

Chapter 5: On God and Beauty

5.1 Introduction

The argument of this book as stated in the introduction is that aspects of the natural science of chemistry as currently practiced, may inform a natural theology. In order to present a coherent argument it has been necessary to present the basis upon which knowledge may be justified in chapter 1, followed by discussions on the particular natural theology I am seeking to make use of, followed by characterisations of contemporary chemical research and concluding with examples taken from recent research papers in chapter 4. The required link between contemporary chemical research and the particular chosen form of natural theology has been the apprehension of beauty in both areas of study. The perception of beauty has been shown to be evident in aspects of contemporary chemical research.

The task in this section therefore is to review historically what has been said on 'God and Beauty' so as to be able to locate our present discussions within the overall Western historical tradition on this subject. I addressed this subject from a scriptural standpoint in chapter 2; here I propose to offer this brief analysis upon the dual bases of the Platonic and the Aristotelian systems.

Any discussion on God and beauty is distinct from reflections on, for example, what beauty is, whether it is objective or subjective, whether it belongs to the object being called beautiful or whether it is purely 'in the eye of the beholder'. It becomes instead an interpretive study as to the particular author's attitude towards estimations of the transcendent. Where the views of several authors are compared, it is easy to portray estimations of beauty as a confusion or a series of dichotomies. As Crispin Sartwell relates, 'it seems senseless to say that beauty has no connection to subjective response or that it is entirely objective' (Sartwell, 2014, p. 3). Interestingly in view of this present book, Sartwell portrays beauty as the result of the relationship between subject and object in which aspects are 'juxtaposed and connected'; it both invites and requires, exploration and inquiry (Sartwell, 2014, p. 12). As a result the perception of beauty would indeed appear to be a lure to seek the Divine, as Richard Viladesau indicates:

> There seems to be no doubt that experiences of beauty can lead the spirit to God and confirm people in devotion, and that therefore the aesthetic dimension is one that must have a place in the communication of religious truths. (Viladesau, 1999, p. 104)

Viladesau is here expressing much of the hope of this book: that an experience or perception of beauty can lead any human spirit *towards* God and that those who are already Christians might be confirmed in their devotions. As a result I contend that it is rational to make use of this mechanism to communicate the truth about God to those who seek Him. I am of course also attempting to take this further in showing that it is specifically the Christian God who is communicated in beauty and in his use of authors such as the Apostle Paul, Karl Barth and Hans Urs von Balthasar amongst others, there is every indication that Viladesau also has this in mind. From the reference to authors so widely spaced in time, it can be seen that any development in views over time is less important than the positions themselves. It is not that there is in any sense unanimity on this question however those positions which are ancient are no less important than those ideas developed more recently.

5.2 The Two Streams

In our contemporary culture, if one were to offer a lecture series with as topic 'On God and Beauty', then in the part entitled 'Plato and Aristotle', those attending the lectures would want to be told precisely what Plato and Aristotle thought about beauty. Why choose these two philosophers? If it is possible to define the inheritance of the major philosophical threads to this debate: 'On God and Beauty', then I offer the premise that two of the most significant threads are based, one on the thinking of Plato and the other on Aristotle. Writers so disparately separated in time and inclination as Pseudo-Dionysius, von Bathasar and even Whitehead inherit broadly from platonic ideals in discussions on beauty whereas of course Aquinas, Kant, Barth and I would be bold to suggest A J Ayer, are more aristotelian in outlook.

The differences in their understandings of 'On God and Beauty' have less to do with what they said or thought, than with how they went about arriving at them. The methodology is, in this specific area of enquiry, which is not to say elsewhere, of greater importance than what their pronouncements were and those of their intellectual descendants, are. Their ways of reaching conclusions differed. Broadly, Plato sought to equip his students and us his readers with the type of questioning mind needed for elucidation, fully recognising that his conclusions were always open to further development. Aristotle, utilising a most powerful intellect and indeed physical abilities as investigator, sought to teach us his ways of enquiry and then to tell us how things actually are, given the results of his extensive work. Both approaches allow for and credit, intellectual rigour and extensive study, but only one approach enables Beauty to be fully beautiful, whereas the other wants to limit or even shun beauty as marginal, enticing but almost corrupting, as a form of emotion rather than an aspect of the factual. One approach fits encounters with Beauty and the beautiful into an overall valid picture of existence and the other keeps it firmly under control lest it overwhelm, confuse and in some sense pollute, a cold intellectual rationality. One expects solid answers and describes it a failure when such are not forthcoming. The other, having established a basis for thinking, is content to question. Much of the thinking on God and Beauty can be validly interpreted through these two aspects or trends or families, of thought.

5.3 The Aristotelian Stream

What writings of Plato we have are in the form of dialogues, never referring to the author's opinions, but forever leaving us with questions. Aristotle's investigations in all manner of disciplines were prodigious and he not only expects to deliver answers, he also gives them. In book 13, part 3, of *The Metaphysics*, Aristotle remarks:

> Now since the good and the beautiful are different (for the former always implies conduct as its subject, while the beautiful is found also in motionless things), those who assert that the mathematical sciences say nothing of the beautiful or the good are in error. For these sciences say and prove a great deal about them; if they do not expressly mention them, but prove attributes which are their results or their definitions, it is not true to say that they tell us nothing about them. The chief forms of beauty are order and symmetry and definiteness, which the mathematical sciences demonstrate in a special degree. And since these (e.g. order and definiteness) are obviously causes of many things, evidently these sciences must treat this sort of causative principle also (i.e. the beautiful) as in some sense a cause. (Aristotle, 2016)

Hence, as Oleg Bychkov explains, it would seem reasonable to understand that according to Aristotle, beauty is something perceived through the senses:

> It is also well known that according to the Aristotelian view on the nature of cognition adopted by Aquinas, the source of all forms for the intellect is sensory perception (Summa Theologica I.84.4) and not the Platonic Ideas, the Neoplatonic celestial "separate substances," or the "agent intellect" if it is understood as a separate substance. The human intellect, according to Aquinas, does not require any special divine illumination. True, God initially gives humans a sufficient capacity for thought, so one can speak of a God-given "natural illumination" or our pre-existing capacity to recognize truth. There is a pre-existing harmony between bodies, the senses that perceive their material forms, and the intellect that is able to abstract immaterial forms from them. Yet, since the intellect is supposed to understand universal natures as existing in particulars, it must turn to the senses. The intellect needs the body for cognition in order to enable the senses to

extract forms from other bodies. Thus the material part of the human being in Aquinas is absolutely essential: according to Aquinas (at least professedly so), the human being is not two things but a unified soul-body compound. (Bychkov, 2014, pp. 3–4)

I am acutely aware that these two quotations from Aristotle and Bychkov, are very widely spaced in time, separated by many centuries. The influences from one to the other will be obvious. Moreover the essential point is that we can by pure reason come to understand what beauty is and how we interact with it. There is no need to consider that beauty might have an origin outside the human condition. Furthermore the contrast between Platonic ideals is also drawn. One looks to the heavens for answers and the other to our intellectual abilities and to the 'earthly' for elucidation. It is surely not too much of a leap to arrive at Immanuel Kant's view which was essentially that beauty was largely a subjective quantity informed by something called 'taste' in the viewer. As he says in chapter 1, 5.203, of the *Critique of the power of judgment:*

> In order to decide whether or not something is beautiful, we do not relate the representation by means of understanding to the object for cognition, but rather relate it by means of the imagination (perhaps combined with the understanding) to the subject and its feeling of pleasure or displeasure. The judgment of taste is therefore not a cognitive judgment, hence not a logical one, but is rather aesthetic, by which is understood one whose determining ground cannot be other than subjective. (Kant, 2000, p. 89)

And again, he adds a note in the same place:

> The definition of taste that is the basis here is that it is the faculty for the judging of the beautiful. But what is required for calling an object beautiful must be discovered by the analysis of judgments of taste.

In 5:204 and 205 Kant further asserts that beauty has to do with desire, pleasure and the like and that it has nothing to do with the 'stuff' of the thing, but has merely to do with what the human thinks of it.

And how does all this come about? It appears reasonable to assume from what Kant tells us in the same volume in chapter 61, 5.359, that this is because humans are simply made that way:

> One has good reason to assume, in accordance with transcendental principles, a subjective purposiveness of nature in its particular laws, for comprehensibility for the human power of judgment and the possibility of the connection of the particular experiences in one system of nature; where among its many products those can also be expected to be possible which, just as if they had actually been designed for our power of judgment, contain a form so specifically suited for it that by means of their variety and unity they serve as it were to strengthen and entertain the mental powers (which are in play in the use of these faculties), and to which one has therefore ascribed the name of beautiful forms. (Kant, 2000, p. 233)

As we have seen elsewhere in this book, that might have been a valid understanding of the beautiful, until beauty quite obviously could be used to predict outcomes in experimentation:

> An aesthetic intersubjective acceptance based on harmony with existing scientific understanding also plays a role, and beauty is a reliable guide for reason in the search for new scientific truths. (Walhout, 2009, p. 774)

From earlier discussions in chapter 4 above as well as here, it can be seen that the contention that beauty has nothing to offer except pleasure, amounts to a reductionist view of what conflations of shape, form and order might do to the psyche of those humans educated in similar appreciations of 'taste' – which together Kant calls 'beautiful forms' -, can be set aside.

Thus in terms of God and the beautiful, such is mere subjectivity under these schemes of thought. It can not be an objective communication from God, as the central thesis of this book requires, because it only relates to physical 'things', has to do with that which causes pleasure, and is ruled by individual 'taste'. As seen in this manner, beauty must be kept at arms-length and be feared.

This manner of viewing beauty, if it were true, has its entirely rational conclusion in the thinking of A J Ayer:

> As we have already said, our conclusions about the nature of ethics apply to aesthetics also. Aesthetic terms are used in exactly the same way as ethical terms. Such aesthetic words as 'beautiful' and 'hideous' are employed, as ethical words are employed, not to make statements of fact, but simply to express certain feelings and evoke a certain response. It follows, as in ethics, that there is no sense in attributing objective validity to aesthetic judgements, and no possibility of arguing about questions of value in aesthetics, but only about questions of fact …. We conclude, therefore, that there is nothing in aesthetics, any more than there is in ethics, to justify the view that it embodies a unique form of knowledge. (Ayer, 1936, p. 118)

It would be interesting to explore in what way the word 'unique' was being used here. Did Ayer consider that he was opposing ethics and beauty as together forming a 'unique form of knowledge' or were each separate (but then surely they could not severally be 'unique'?). No matter: his sense is clear, considerations of beauty specifically have to do with emotion and emotion can not lead to 'fact' or considerations of value. Yet he reaches this conclusion only because he has already decided that beauty as a concept has little to do with generating truth.

Perhaps astonishingly yet similarly Karl Barth also cautions his readers that beauty is neither 'a leading concept' nor a 'primary motif in our understanding of the whole being of God' (Barth, 1957, p. 652). Yet also in a somewhat similar and I would venture aristotelian championing of human capability, he decries any action that would lead to 'a blind spot in our knowledge' of God (Barth, 1957, p. 650). Barth tells us that God's trinitarian nature is the basis of His power and dignity and the 'secret of His beauty' (Barth, 1957, p. 662). Reading this, it appears rational to say that Barth is attesting to God being beautiful on account of His being Triune and that a *communication* of God's beauty arises on account of His glory, power and dignity – but that beauty is not a leading concept! Moreover Barth is quite rightly fearful lest an aestheticism should come 'to have and keep the last word' as it would become an idol, yet it is not obvious why anyone reading his text and reveling in God's beauty would or could rationally uncouple the aesthetic from the Divine which feeds it. As a result God's beauty could not possibly be construed as 'the last word' and is instead something that follows from who He is. Talk of 'last words' in Barth's texts does have echoes of his objections to natural theology: Barth does not want to detract in any way from God's self-revelation of Himself and neither would I as God's Beauty can not be a stand-alone concept separate from the God who projects it.

Barth's tortured reasoning is underlined when he says in the same place that were God to loose dignity and the 'power of real divinity' God would no longer be beautiful – but beauty is still not a leading concept in our knowledge and understanding of God! Far from not being a leading concept I would affirm in contrast that to experience true beauty is indeed to have had a foretaste of the full presence of God, which seems apophatically to be precisely what Barth is implying but steadfastly refusing to say. Thus a sensation of a 'something' being beautiful acts as a signifier of the presence of God being communicated to us. As a light shining from a doorway is a comfort to a weary traveller on a lonely road to the effect that their lodging place for the night is not far off, so too a perception of the beautiful tells us that Someone else is, in making themselves known, open to being communicated with.

In a highpoint of Christological logic, Barth goes on to say that 'in the name and person of Jesus Christ' the beauty of God is revealed 'in a special way and in some sense to a supreme degree' and that we only know this to be so 'from the existence of the Son of God in His union with humanity' (Barth, 1957, p. 662). Barth strongly implies through his use of quotations that his understandings on these points are informed by Aquinas (*Summa Theologica I*) which observation completes the aristotelian strand of thinking on God and beauty, in making it somewhat circular.

5.4 The Platonic Stream

Yet let us stay with Barth for just a little while longer as we now explore the platonic strand to thinking on God and beauty. He accuses both Augustine and Pseudo-Dionysius of expounding 'the beauty of God as the ultimate cause producing and moving all things' (Barth, 1957, p. 652) but as we shall see in the case of Pseudo-Dionysius, what Barth has succumbed to is his insistence on not having discrepancies in his knowledge of God. Such knowledge *pace Barth,* is in fact open-ended, hidden, dark, mysterious and in Christ radically transforming – overturning and reversing – of our earthly concepts so that in Him we can expect to 'take every thought captive to the obedience of Christ' (2 Corinthians 10.5) whilst similarly being un-knowing of 'what we will become' (1 John 3.2–3). In order to approach the un-approachable God we must relinquish any insistence that humans can probe all mystery. Having mentioned Pseudo-Dionysius and a potential misunderstanding of his thought it is to him we must now turn.

In surveying discussions of God and Beauty what is communicated, notwithstanding the comments on Barth above as we shall see, is a certain awe in the face of God's beauty, as Pseudo-Dionysius recounts in *On the Divine Names* chapter 4 section 13:

> And in truth, it must be said too that the very cause of the universe in the beautiful, good superabundance of his [God's] benign yearning for all is also carried outside of himself in the loving care he has for everything. (Pseudo-Dionysius, 1987, p. 82)

He goes on to say in the same chapter and section 14: 'for in the end what is he if not Beauty and Goodness' (Pseudo Dionysius, 1987, p. 82).

Compare this from Barth:

> In this self-declaration, – the unity and differentiation – however, God's beauty embraces death as well as life, fear as well as joy, what we might call the ugly as well as what we might call the beautiful. It reveals itself and wills to be known on the road from the one to the other, in the turning from the self-humiliation of God for the benefit of man to the exaltation of man by God and to God. (Barth, 1957, p. 665)

And again from Augustine in his *The City of God,* Book XI, Chapter XVIII:

> For God would never have foreknown vice in any work of His, angel or man, but that He knew in like manner what good use to put it unto, so making the world's course, like a fair poem, more gracious by antithetic figures. *Antitheta,* called in Latin opposites, are the most elegant figures of all elocution: some, more expressly, call them contra-posites. ..... Thus as these contraries opposed do give the saying an excellent grace, so is the world's beauty composed of contrarieties, not in figure but in nature. (Augustine, 1973, p. 327)

By re-using the thinking alluded to by Augustine in the same place utilising words from the Biblical Apocryphal book *Ecclesiasticus* ('against evil is good, and against death is life'), there is at least the suspicion that Barth consulted both. The beauty that is revealed through God in Christ is shown to be arresting, in a series of radical opposites.

As a result Kevin Corrigan and L. Michael Harrington speak of Pseudo-Dionysius feeling the need 'to push language forms to their breaking points, and then to see what we cannot say about God' (Corrigan and Harrington, 2015, p. 22). Pseudo-Dionysius makes use of such opposites as signifiers, that lead our human linguistic consciousness to entertain the otherness of God through the forced consideration of words about God which are plainly almost absurd (Corrigan and Harrington, 2015, p. 10). In this way and amongst many such examples Pseudo-Dionysius says that God is both nameless and 'has the names of everything that is' (Pseudo-Dionysius, 1987, p. 56). As a result we can validly subscribe to a radical 'open-endedness' about our knowledge of, and the possibility of encounters with, God, where for example to say with any sense of finality that God 'is' something, is to in effect shut-off further exploration and meditation. They are telling us that this makes Pseudo-Dionysius a dangerous writer but I think this is taking matters in the wrong direction: Pseudo-Dionysius upholds a strong Christology with profound insights into the Trinity and enjoins a strict liturgical practice (Corrigan and Harrington, 2015, p. 27). He insists that we must 'use only what scripture has disclosed' when using words of the 'hidden transcendent God', that is, when thinking theologically (Pseudo-Dionysius, 1987, p. 50). As concerns the beauty of God, these authors remark that Pseudo-Dionysius' view of the 'visible created Universe' upheld a 'vivid sense of the aesthetic and imaginative beauty of the sensible universe, pervaded from the perspective of divine beauty by interrelatedness and harmony' (Corrigan and Harrington, 2015, p. 28 and see also Pseudo-Dionysius, 1987, pp. 54, 55). Thus whereas Barth speaks of warnings and fear in our embracing of God's beauty, Pseudo-Dionysius looks for it everywhere and exhibits no fear embracing it, because God is all and in all (Colossians 3.11).

Pseudo-Dionysius takes us on a form of journey into an ever-closer relationship with God in Christ, unbounded by our preconceptions and yet retaining the orthodox Trinitarian Creedal formulations and wedded to traditional liturgical practices and scriptural interpretations. [Corrigan and Harrington re-interpret or extend Pseudo-Dionysius' writings at this point to apparently permit inter-religious dialogue, but this amounts to a contemporary extension rather than a re-interpretation of his writings. Yet upon Pseudo-Dionysius' words alone, his theology enables a divine reading and re-reading of both nature and word, laying us open to God's sacred works and indeed Person (Corrigan and Harrington, 2015, p. 26)].

And it is into such a radically new landscape that true beauty transports us, a translation that 'belongs to the very origin of Christianity' and through which 'Jesus' figure stands out in his encounters and conversations', and as a result of which 'the Unconditional breaks through, casting a person down to adoration and transforming him into a believer and follower' (von Balthasar, 1982, p. 33). Beauty as a sudden interruption to the ordinary and the every-day has such force that when in combination with talk of the radical Other that is God, could manifest itself as either ecstasy or ruin depending on the observer's view of reality: ecstasy for the person suddenly finding themselves in God's presence and ruin for any thought that stands against the Divine:

> Both the person who is transported by natural beauty and the one snatched up by the beauty Christ must appear to the world to be fools, and the world will attempt to explain their state in terms of psychological of even physiological laws (Acts 2.13). But *they* know what they have seen, and they care not one farthing what people may say. They suffer because of their love, and it is only the fact that they have been inflamed by the most sublime of beauties – a beauty crowned with thorns and crucified – that justifies their sharing in that suffering. (von Balthasar, 1982, p. 33)

Let the reader note how here von Balthasar is speaking of potentially two groups of people both being transported by beauty: the one (possibly non-Christian) being moved by the natural world, and the other (Christians, no-doubt through Divine revelation) being 'snatched-up' by Christ's beauty. Von Balthasar speaks of both groups being 'inflamed' by the one beauty which is that 'crowned with thorns and crucified'. We are then urged, against Barth's warning, to cross:

> the boundary between the realm of nature and that of grace .... The form of the beautiful appear[ing] to us to be so transcendent in itself that it glided with perfect continuity from the natural into the supernatural world. (von Balthasar, 1982, p. 34)

The role of the beautiful here is being shown to be powerfully transcendent in mechanism: it is as though the sight of the observer is being drawn away from the sight of the natural world and its beauties, towards Christ as its source. This is plainly at odds with the attitudes expressed by for example Barth or indeed by McGrath elsewhere in this book. How are we to resolve these apparent tensions? Does not the resolution lie in the *effect* the encounter with the objective (a perception of beauty in this case), has on the subject? If and when God in Christ, encountered as beauty in the natural world does indeed lead to the person coming to Christ, becoming a Christian, then surely it would be deeply wrong to deny them entrance into the community of the faithful (see Acts 10.47)? And when this world seeks to subvert the beauty of Christ revealed in the world that He made, perhaps in a form of alluring yet ultimately perverting aestheticism that

Barth warns against, then we 'suddenly come to an astonished halt and conscientiously decline to continue on that path' (von Balthasar, 1982, p. 37).

The reader will now appreciate the mechanism of the beautiful if read in the platonic sense, as representing a truer reality than the world or earthly concreteness on display. God's nature, so closely bound to God's beauty, gives the force, the motivating power behind, in truth the love, which can so radically transform the individual who encounters Christ.

As Whitehead himself remarks, it is this seeking of the 'forms in the facts' within Platonic philosophy's 'abiding appeal' that allows his Process philosophy to become one of the more radical expressions of Platonism (Whitehead, 1978, p. 20). With this methodology then, Whitehead's schema acknowledges the transcendental:

> if we had to render Plato's general point of view with the least changes made necessary by the intervening two thousand years of human experience in social organization, in aesthetic attainments, in science, and in religion, we should have to set about the construction of a philosophy of organism. In such a philosophy the actualities constituting the process of the world are conceived as exemplifying the ingression (or 'participation') of other things which constitute the potentialities of definiteness for any actual existence. The things which are temporal arise by their participation in the things which are eternal. (Whitehead, 1978, pp. 39–40)

From this it is clear that in a Process Theological view, beauty perceived in a temporal sense can indeed be a signifier of the eternal beauty in God, owing to the beauty we can see participating in the eternal, which we can not.

Any tensions, for that is what I would suggest they are, between an orthodox Christian Faith and a Process Theology, are indeed recognised to exist in von Balthasar's writings, as Gerard O'Hanlon observes:

> The challenge to the traditional axiom of divine immutability in Process Theology is based on the philosophical principle which affirms the primacy of becoming over being. It is fascinating to note that the issue arises in Balthasar in a very definitely theological context, and moreover within a Christology and theology 'from above' in which the philosophical component is respected and given its due but in which the theological retains a certain priority and normativity. (O'Hanlon, 1990, pp. 10–11)

Yet these tensions have already been collapsed in the person of Christ:

> The Son, then, as Word of the Father reveals and expresses the Father. More specifically, too, it is especially through the obedience of his life and death that the Son carried out this saving revelatory role .... the high point of this obedience lies in the event of the cross .... and so it is this event, at the centre of the revealing and dynamic 'figure' of Christ, which is also at the centre of Balthasar's theology and from which all else is interpreted. The cross is the exposed place in which love appears at its most extreme and as most itself. (O'Hanlon, 1990, p. 10)

In summary then, it is possible to recognise some helpful elements in Whitehead's approach to God and Beauty, most notably the recognition of an eternal truth underpinning in some manner, that which is perceived in the temporal. Our gaze nonetheless, as O' Hanlon reminds us, must remain centred on the figure of Christ. Beauty is not merely indicative of a Whiteheadean underlining eternal 'thing' but reveals the dynamic figure of the Christ.

5.5 Conclusion

For well over 2000 years a person's attitude to perceptions of beauty has provided a window into their willingness to entertain variously:

- the possibility of the transcendent breaking into their temporal existence
- whether one is content to live with and within mystery and wonder (including that mystery and wonder uncovered through scientific research) or whether in contrast one holds that everything in the physical world must be both explained and explicable, and have a purely temporal explanation
- One can live without notions of beauty clouding one's rational vision yet it is likely to be a poor interpretation of the complete reality. That there is a something called 'beauty' that can captivate, stun or even transport those who are willing to see it for what it is, is an assertion perhaps many will concede. That it provides sight of the Christ who saves, is in God's gift by grace alone, but what is certain is that the potentiality of that soteriological witness, is within the grasp of all who choose to reach out.
- It may however be possible to leave this brief thematic survey of 'God and Beauty' with a more helpful synthesis of these two positions or streams. Perhaps I have drawn the lines too starkly: Aquinas as an interpreter of Aristotle is almost certainly one of the greatest of Christian writers and whose analyses are frequently still being turned to today whereas in contrast Whitehead, who considered himself a platonist, was not an orthodox Christian: truth appears to lie in both positions. Earlier I remarked how Derkse tells us that science has the task of *decomposition* for the purposes of comprehension and *recomposition* in order to make use of it all. In the same place I reported how Bulkeley implies that a certain destructiveness accompanies that decomposition which is then transformed into a *renewed* and *expanded* 'capacity for surprise amazement and curiosity' in this recomposition *phase* of any sustained encounter with the detail of the natural world. Perhaps this then is a more helpful way of seeing the flow of historical comment on

'God and Beauty': by all means dive into the detail and attempt to explain as much as you can, but always remember to come back to an overview. The plain and simple fact is the wonder of it all: what God has made is indeed very, very beautiful.

Chapter 6: Discussion

6.1 Introduction

The objective of this book is to show that chemistry can indeed rationally inform a natural theology. In this final chapter I will draw together the arguments mapped-out in the preceding chapters to demonstrate how this might be achieved. The discussion has proceeded as follows: it was first necessary to establish an epistemological position on the basis of which it is rational to believe in the Christian God. Following this I explored natural theology as a mechanism for demonstrating this rationality to those who are not Christians. Within this second chapter I also developed a particular form of natural theology that I proposed was a best 'fit' for the discipline of chemistry. In the third chapter I familiarised the reader with aspects of chemistry both historical and contemporary. This was expanded to reveal the metaphysical in chemistry. I also began to show how beauty can act as a bridge between the natural theology and the practice of chemical research. A closer inspection of beauty was here also necessary to show how aspects of its perception are specifically attuned to the ways in which results of chemical research are seen by practitioners. A small survey of contemporary research papers emphasised those aspects of chemistry and beauty already alluded to. A discussion of God and beauty in the fifth chapter, in addition to illustrating how God has been shown to be related to beauty, also detailed how the obvious beauty present in creation can and is ignored by commentators: an openness to beauty is required in order to perceive the Divine in creation.

We are now almost ready to draw matters together. There remains a gap in the discourse which needs to be resolved and this concerns beauty itself. Whilst I have written in some detail on the subject in both theological and chemical terms, what has been lacking up-to-now has been a definition: how can I make use of the perception of beauty as both bridge between a natural theology and chemistry and as a pointer towards Christ in creation if I can not say what it is?

6.2 Defining Beauty

In finally being able to come to a definition of beauty that not only encompasses forms of art and music but also more modern pursuits such as computer software creation and most importantly here the efforts in and products of, chemical research, I am in part reliant upon Wynn's statement that:

> ... God is not simply a powerful individual whose purposes are good, but a uniquely concentrated expression of what it is to be. On this view, the goodness and beauty of the world provide a clue not just to God's benevolent intentions in relation to the world, but to the goodness and beauty of the divine being itself.... If asked what the world basically is, our reply [should be] 'a locus of value'. (Wynn, 1999, p. 196)

Importantly Wynn does not say that the beauty of the world is or equals the beauty of God, but that it provides *clues* to the beauty of God. He speaks of God as 'Beauty and Meaning and Love' (Wynn, 1999, p. 186). Having earlier in his book explored notions of resemblance 'in order to spell out the sense in which the world represents God' Wynn then goes on to explain the nature of the relationship between the 'thing' that is perceived to be beautiful and its ultimate source in God as one of *complementarity* rather than resemblance (Wynn, 1999, p. 179). Wynn explains his use of complementarity as follows:

> ... we may suppose that if one part of such a work [he is speaking of a work of art] were to be removed, then the character of that part could in principle be inferred from a knowledge of the remainder of the work .... We sometimes suppose that there is one thing and one thing only which is able to complete a work of art. Let us call the relationship which binds one part of a work of art to the thing which is able uniquely to complete that part the relationship of complementarity. (Wynn, 1999, pp. 179,180)

Crucially for the definition which I propose below, Wynn implies a form of separation between the beauty we see and the God who is its source, whilst maintaining the connection in complementarity:

> .... one part of an aesthetic object may represent the rest of the object. Notice that in such cases, the representation does not turn upon resemblance: there is no necessity that the complementary element should mirror (or be mirrored by) the element which is already in place. Nor is the representation like the merely conventional representation which is characteristic of linguistic denotation. Nor yet is it like the relation of symptom to cause. Instead, it has to do with the way in which one object may uniquely identify another by virtue of the aesthetic relationship which unites it to this further object. (Wynn, 1999, p. 180)

In seeking a rich exposition of complementarity, I turn now to a schema prepared by Richard Kearney which 'attests to ways in which the sacred is in the world but not of the world', the sacred inhabiting the secular but not being identical with it, although crucially both needing each other, as our description above from Wynn affirms (Kearney, 2011, p. 152). And such phrases are plainly reminiscent of Christ's command to be in the world yet not of it (John 17.14–18; see also Romans 12.2).

In illustrating this notion, Kearney speaking of the moments throughout history when people make a break with 'ingrained habits of thought and open up novel possibilities of meaning' and in so doing suspend 'received assumptions' to enable us to be 'open to the birth of the new' (Kearney, 2011, p. 7), proposes three as he says 'arcs' or perhaps more meaningfully, components of such moments. These he terms the iconoclastic, the prophetic and the sacramental (Kearney, 2011, p. 152). Plainly Kearney is not thinking of chemistry being brought into a natural theology and yet for the illustration of the idea of complementarity in the implementation of new ways of thinking about beauty, his text is helpful. In this way, iconoclasm 'unmasks mendacious and illusory idols' and stands in protest against them (Kearney, 2011, p. 153). This I would suggest, accurately reveals the manner in which a purely aesthetic way of life, beauty for beauty's sake without thought to its origin, is ultimately ruinous. This I would contrast with my account of beauty which, in revealing Christ as its source, is life enhancing and affirming. Secondly the prophetic component 'lets symbols speak of new things still to come', as Kearney says in the same place, in an 'hermeneutics of reaffirmation', gaining back 'a living God after forsaking an illusory one'. This in my present book speaks powerfully in favour of the use of models and other symbology in the building of a critically realist view of the often unseen realities, that lie at the heart of much chemical exploration and indeed certain Christian theological mysteries. It rejects an insistence on a dogmatic fixed order of things as though this were more comforting and instead embraces an 'ontological openness' as Polkinghorne expresses it, quoted in chapter 3 above. Thirdly and finally the sacramental which is recovered in the 'lived' world of suffering and action', 'complements [this] prophecy of promise with concrete attention to embodied divinity' (Kearney, 2011, p. 153).

Thus in his use of a certain parallelism to equate his earlier 'iconoclasm' with 'concrete attention to embodied divinity' and in his use of 'the prophetic' paralleled with a 'prophecy of promise', Kearney's treatment can be extended to include:

a recognition of what he calls – perhaps inelegantly – 'sacred enfleshment' (Kearney, 2011, p. 7), which I take to mean concrete examples of what Paul is speaking of when he talks of God's eternal power and divine nature being clearly seen in what has been made (Romans 1.18,19) – embodied divinity indeed;

a certain humility in the face of all we do not know and therefore having recognised the 'indispensable significance of a moment of dispossessive bewilderment', to 'surrender inherited sureties and turn towards the Other – in wonder and bewilderment, in fear and trembling, in fascination and awe' (Kearney, 2011, pp. 8,11); none of which sounds very far from the emotions expressed in the discoveries made and reported upon in chapter 4 above and elsewhere, and accurately expresses the hope, the prophetic promise, of the conversation that lies at the heart of this book.

Thus I have extended Kearney's exposition to propose ways in which God in Christ can break through those previous 'certainties' of old, perhaps certainties on both sides of the argument: those of scientific or naturalistic positivism on the one hand and religious dogma on the other (or as Kearney puts it, a way 'that precedes and exceeds the extremes of dogmatic theism and militant atheism' 2011, p. 166). And so finally to borrow from Kearney and to extend his use of understandings of the secular and sacred: I am not saying that complementarity means that the beauty we see in nature is God yet rather that God is visible in and through the beauty perceived in the natural, and crucially that it is bound-up with the natural world and that the natural world is thus pointing towards God ('the sacred is [not] the secular; … it is in the secular, through the secular, toward the secular. I would even go so far as to say the sacred is inseparable from the secular, while remaining distinct': Kearney, 2011, p. 166). And in being inseparable yet distinct as described, such an explanation of complementarity neatly chimes with that of Wynn above.

Returning now to Wynn: he does not give a definition of beauty but in doing so here I make use of this notion of complementarity in order to achieve several outcomes:

- I am proposing that the beauty being perceived is part of the mechanism of the particular argument from design that I am advocating. As part of a natural theology, as I have already stated, it can not prove God's existence but merely indicate or point at, the Divinity. Beauty functioning as 'complementary' maintains the duality I spoke of in chapter 2 above whilst still being linked to its source in the Creator.
- In line with the completion of the account of beauty offered in section 5.2 above I hold that love of the 'thing' that is perceived to be beautiful forms part of that apprehension of beauty. Yet I am not convinced that I 'love' the things a molecule is made from or consists of, like the example of Buckminster Fullerene given above in chapter 4, in the sense of being 'in love' with the carbon from which it is made. There is something I am loving as a result of my finding it beautiful, but I don't 'love' a chemical. Again beauty functioning as 'complementary' to the 'thing' found to be beautiful is a more satisfactory resolution to the problem of loving the 'thing' itself.

- In chapter 2 I discussed what it was that invoked the notion of beauty as expressed for example in Psalm 73 and elsewhere, and came to the conclusion that a perception of beauty arose as a result of understanding that one was in God's Presence. It was not that one enjoys visiting a spectacular (religious) building or appreciates the interior decor or fittings, but that one had been gifted the sense of being in the Divine Presence. Beauty functioning as complementary or '*in complementarity*', again here resolves the potential confusion of being in love with or of finding beauty-in, a 'thing', for example a building, as opposed to the Divine who made Himself present to the person within that building or space. Thus the given definition addresses the tension at the heart of beauty as both 'event' or 'being', to use Westermann's terminology.

I should now like the reader to consider the multi-vocal term 'shape'. We might speak for example of a statue or any solid physical object as having a 'shape' in three dimensional space. In being crafted it is usual to speak of 'shaping' an object. In musical performance it is usual to also use shape as a verb in speaking of 'shaping a phrase' in the music. A chemical compound has a shape both as a two dimensional diagrammatic representation and as a three dimensional form in space even though it cannot be seen. Although not commonly used, if I were to speak of the shape as the design of a piece of software, that would be largely intelligible even though plainly I am not speaking of the shape of the typeface on the computer screen but the manner in which the functionality has been assembled. In a painting or any other form of image I could quite readily speak about the shape of for example a face or of clothing even though it is only being rendered in two dimensions. One might speak of a person's emotions for example shaping an interaction or indeed being shaped through an interaction. Note how if for example I discover that I find a particular statue beautiful, I am responding not to a love of marble but rather to what my interaction with, and interpretation of, the surface of that marble *effects in me*. I am therefore interacting with something that is real, that I am envisioning as being beautiful.

Thus overall the word 'shape' speaks about an envisioned surface with which the observer interacts. This surface might be static as in the plastic arts or dynamic as in music and dance; it may be physically solid or perceived in the mind. When used in this manner, 'shape' does indeed speak of something real in the sense of some event exterior to the person having given rise to the perception of the surface: we are not speaking about an object seen in a dream. If this multi-vocal use of the word 'shape' is accepted I offer the following definition of beauty whilst keeping in mind Wynn's use of the notion of complementarity discussed above:

Beauty is the perception of a shape as invoking sacred wonder: a shape complementary to a subject who holds it in value and with love.

The reader might note some elements of this definition before I go on to demonstrate its use:

- the definition is plainly in two parts;
- a perception of beauty extends beyond the first part of the definition, which only relates to the pleasing shape *in* and *of* itself;
- 'sacred' is used in a manner consistent with Kearney's usage of 'sacred' as quoted above: it is that aspect of a process or object or an event which though plainly part of the physical world nonetheless appears unique and in some manner separate from the mundane and everyday;
- 'wonder' as used in the definition, is that aspect of a process or object or an event which causes the person perceiving it to halt and gaze as though unsure of the ultimate provenance of the sensation being perceived;
- the definition is extended into the second part to include the subject that the beauty in complementarity belongs to and without which it would not be complete. In so doing the definition recognises the transcendental nature of the perception of beauty;
- the shape and the act of perception are both dynamic elements of the definition: the shape may only exist fleetingly and the beauty of it is only perceived actively from moment to moment;
- the subject which in complementarity projects or is responsible for, the perceived beauty, is in a Christian understanding Christ in creation. Where Christ is not understood in this way, it is that transcendent aspect of humanity which creates the wonder and mystery frequently, as I have explained above, felt by for example lovers of music or chemical researchers;
- the 'locus of value', to re-use Wynn's terminology, is expressed by the subject;
- plainly, in this definition, removal of the subject would remove the perception of beauty.

6.3 Conclusion

I shall now re-visit certain key points in the overall argument to check that the given definition of beauty is fit for purpose. Theologically I have moved my version of the design argument as expressed in Chapter 2 away from an appreciation of Creation as strictly ordered or as Wynn says 'lifeless and machine-like', towards one which 'may be grounded in an evaluatively rich conception of the world' (Wynn, 1999, p. 196). Such a more complex appreciation of the character of Creation accepts that its development is contingent and entertains a greater level of in-determinedness in outcomes. In its use of the frameworks set within Creation, chemistry should not be seen as 'unnatural'; those who make things do so according to rules and laws which the Creator put in place. In this manner, overall the character of the natural theology being used is revealed to be aligned with the everyday experience of life.

The definition of beauty given above allows it to be acknowledged as such in all manner of disciplines and in all manner of endeavours, including as was discovered in chapters 3 and 4, in chemistry. The appreciation of beauty was then expanded upon in this present chapter. The given definition is well able to encompass the challenge of the unseen beauty within chemistry. In this way, beauty might be perceived variously for example in the elegant design of a chemical process, in the design of a particular molecule or in the unfolding of a principle from the particular to the general. Examples of each of these are found in chapters 3 and 4 above. Thus the perceiving of beauty in aspects of chemical research is an accepted part of the discipline.

Also in Chapter 2 and then affirmed again in chapter 3, I showed that the ultimate source of beauty is Christ through whom the world was Created. I asserted that the wonder and sense of beauty perceived in the artefacts of the world arises quite naturally from Christ who gave them existence. The world was created through Christ and this world is a 'locus of value' as Wynn remarks: the beauty perceived in the world arises as a result of the value Christ in love places in it. It is thus rational to conceive of such value as having some aspect of complementarity to Creation. Through the work of Wynn and in consideration of the definition of beauty, it has become possible to rationally provide clues to link the beauty perceived in Creation, in the natural world, with the Divine agency through whom it was brought into being. Again the beauty that is Christ, partly hidden yet still shining through the fabric of Creation, requires a definition that can be powerfully, realistically and convincingly present in the mind of the beholder whilst still not be present as evidence for His existence. In this manner the duality in the presentation of the natural theology is maintained: both the natural theologian and the chemist perceive the beauty present in the result of the chemical research. The Christian draws from it encouragement for their lived-in Christian Faith and the non-Christian is left wondering at the Source. Once again the offered definition of beauty appears to be able to maintain such a duality and keeps the beauty discerned as actual and real.

The given definition of beauty insists as does Wynn, that the world is this locus of value since beauty is found in the world and that beauty is itself an acknowledgement of inherent value in the 'shape' that is discerned to be beautiful.

Again I reiterate the question at the heart of this book: can chemistry inform a natural theology? Chemistry has been shown to have a metaphysical quality; chemical research and its products, can be seen to be entirely natural; a transformational ethos in the preparation of new compounds, does not preclude an appreciation of the laws and rules embedded within the natural world. A lived-in or lively Christian faith does not deny the role of evidence in the elucidation of ever-greater mysteries within Creation: indeed it encourages it. The challenge of increased and increasing complexity and knowledge of and about the natural world, does not threaten such a revelation of God to the individual: instead it recognises that knowledge is always 'in part' and our understanding of it 'in part' (1 Corinthians 13.12).

Humans have an ability to place the form of simplicity over the reality of complexity and in so doing forge models and ways in which to advance knowledge for the good of all. This forging creates opportunities within research to discover an ever greater incidence of plainly wondrous artefacts and intimations of this 'beauty beyond our imagining'.

As presented in this book there is ample evidence for the perception of the beauty of God theologically as well as the perception of beauty within certain aspects of chemical research. In relying on Plantingan warrant to convert belief into knowledge as expounded in chapter 1, the epistemological position put forward here privileges these 'perceptions of beauty' in both theology and chemistry as knowledge. In theological terms this knowledge amounts to an acceptance of Christ as the source of that beauty. In matters of chemistry such knowledge credits beauty as an objective reality and as an aide to further advances in research. The perceptions of the beautiful in chemistry have the same source as beauty in other disciplines although in a natural theological sense are presented as mysterious and 'other'.

The particular form of critical realism proposed in this book, hints of links to such perceptions of beauty with the Creator who shaped it. A form of the Argument from Design as the particular vehicle for the natural theology being used here, calls after Wynn's and Kearney's insights, for a rich holistic appreciation of Creation that at its core exhibits beauty in many and varied ways, as a complementary aspect of it being crafted by its Creator. Such *complementarity* in the artifice of Creation does not allow for exact detailed predictions as to the outcomes of all research into the workings of the natural world yet at every turn does reveal, for those who as we discovered in Chapter 5 are willing to entertain it, the hand of the Creator in all its beauty. Chemistry can indeed inform a natural theology.

Bibliography

Amesbury, R., 2014. Fideism. *The Stanford Encyclopedia of Philosophy* (Winter 2012 Edition), E. N. Zalta (ed.), URL = <http://plato.stanford.edu/archives/win2012/entries/fideism/> [Accessed 31 October 2016].

Aristotle, 2016. Metaphysics, Book 13.[online]. Available at: <http://classics.mit.edu/Aristotle/metaphysics.13.xiii.html > [Accessed 22 July 2016].

Augustine, and Healey, J., 1973. *The City of God, Volume 1*. London: Dent.

Ayer, A.J., 1936. *Language, Truth and Logic*. Reprint 2001. London: Penguin Books Ltd.

Bagger, M., 1999. *Religious Experience, Justification, and History*. Cambridge: CUP.

Baker, D., 2007. *Tayloring Reformed Epistemology: Charles Taylor, Alvin Plantinga and the De Jure Challenge to Christian Belief*. London: SCM Press.

Barbour, I.G., 1990. *Religion in an Age of Science*. London: SCM Press.

Barbour, I.G., 2013. *Myths, Models and Paradigms*. [Kindle version] HarperOne. Available at: Amazon.co.uk <http:// www.amazon.co.uk> [Accessed 22 January 2016].

Barr, J., 1993. *Biblical Faith and Natural Theology: The Gifford Lectures for 1991, Delivered in the University of Edinburgh*. Oxford: Clarendon Press.

Barrett, J.L., 2012. Natural Theology after Modernism. In: J.B. Stump and A.G. Padgett, eds 2012. *The Blackwell Companion to Science and Christianity*. Chichester: Wiley-Blackwell. pp. 319–334.

Barrow, J.D., 2007. *New Theories of Everything: The Quest for Ultimate Explanation*. New York: OUP.

Barth, K., 1946. No!. In: *Natural Theology: Comprising 'Nature and Grace' by Professor Dr. Emil Brunner and the reply 'No!' by Dr. Karl Barth*. Translated from German by P. Fraenkel. London: The Centenary Press.

Barth, K., 1957. *Church Dogmatics, Volume 2: The Doctrine of God, Part 1*. Translated from German by T.H.L Parker, W.B. Johnston, H. Knight, J.L.M. Hair. Edinburgh, T. & T. Clark.

Bengoetxea, J., Todt, O. and Luján, J., 2014. Similarity and Representation in Chemical Knowledge Practices. *Foundations of Chemistry*, 16 (3), pp. 215–233.

Bennett-Hunter, G., 2014. *Ineffability and Religious Experience*. London: Pickering & Chatto.

Berger, R., Wagner, M., Feng, X. and Müllen, K., 2015. Polycyclic aromatic azomethine ylides: a unique entry to extended polycyclic heteroaromatics. Chem.

Sci., 2015(6), p. 436. Available at: http://doi.org/10.1039/c4sc02793k [Accessed 8 January 2015].

Berlin, A. and Brettler, M. Z. eds., 2004. *The Jewish Study Bible*. New York: OUP.

Bhaskar, R., 1997. *A Realist Theory of Science*. 2nd ed. London: Verso.

Boyle, R., 1660. *Various Arguments Against Atheists and in Favour of God's Existence*. Available at: <http://www.bbk.ac.uk/boyle/boyle_papers/bp02_docs/bp02_057v-058r.htm> [Accessed 4 April 2014].

Brooke, J.H., 2010. Science and Secularization. In: P. Harrison, ed. 2010. *The Cambridge Companion to Science and Religion*. Cambridge: CUP. p. 103.

Brooke, J.H., and Cantor, G.N., 2000. *Reconstructing Nature: The Engagement of Science and Religion*. Edinburgh: T & T Clark.

Brown, R. E., Fitzmyer, J. A. and Murphy, R. E. eds., 1990. *The New Jerome Biblical Commentary*. London: Geoffrey Chapman.

Brümmer, V., 1981. *Theology and Philosophical Inquiry: An Introduction*. London: The Macmillan Press Ltd.

Brümmer, V., 1992. *Wijsgerige theologie in beweging: Een selectie uit de essays van Vincent Brümmer*. Franeker: T. Wever B.V.

Bulkeley, K., 2005. *The Wondering Brain: Thinking about Religion with and beyond Cognitive Neuroscience*. New York: Routledge.

Burnett, R.E., 2013. *The Westminster Handbook to Karl Barth*. Louisville: Westminster John Knox Press.

Bychkov, O.V., 2015. Metaphysics as Aesthetics: Aquinas' Metaphysics in Present-day Theological Aesthetics. *Modern Theology*, 31(1), pp. 147–148.

Cargile, J., 1992. Pascal's Wager. In: R. D. Geivett and B. Sweetman, eds. 1992. *Contemporary Perspectives on Religious Epistemology*. Oxford: OUP. pp. 283–289.

Cartwright, N., 1983. *How the Laws of Physics Lie*. Oxford: OUP.

Casserley, J.V.L., 1955. *Graceful Reason: The Contribution of Reason to Theology*. London: Longmans, Green and Co. Ltd.

Chapman, A., 2008. From Alchemy to Airpumps: The Foundations of Oxford Chemistry to 1700. In: R.J.P. Williams, J.S. Rowlinson and A. Chapman, eds., 2008. *Chemistry at Oxford: A History from 1600 to 2005*. Cambridge: RSC Publishing. pp. 17–51.

Collins, F., 2009. Learning the Language of God. In: R. Bancewicz, ed. 2009. *Test of Faith*. Milton Keynes: Authentic Media. pp. 1–10.

Cordonnier, M.A., Kan, S.B., Jennifer, G., Birgit, G., Shermin, S. and Anderson, E.A., 2014. Carbopalladation of bromoene-alkynylsilanes: mechanistic insights and synthesis of fused-ring bicyclic silanes and phenols. *Org.*

Chem. Front., 2014(1), pp. 661–673. Available at: <http://doi.org/10.1039/C4QO00123K> [Accessed 10 January 2015].

Corrigan, K. and Harrington, L. M., 2015. Pseudo-Dionysius the Areopagite. *The Stanford Encyclopedia of Philosophy* (Spring 2015 Edition), E. N. Zalta (ed.), URL = <http://plato.stanford.edu/archives/spr2015/entries/pseudo-dionysius-areopagite/> [Accessed 31 October 2016].

Dahling-Sander, C., 1999. Karl Barth – Emil Brunner, An Uneasy Correspondence from the Very Beginning. In: *Karl Barth Archiv*, 2, pp. 8–14. Available at: <https://karlbarth.unibas.ch/fileadmin/downloads/letter2.pdf> [Accessed 25th April 2015].

Dattatraya, H.D., Balu, D.D. and Raghavender, B., 2015. Hg(OAc)2 mediated highly regio- and/or diastereoselective allylic tert-acetylation of olefins. *Org. Chem. Front.*, 2015(2), p. 159. Available at: <http://dx.doi.org/DOI:10.1039/c4qo00310a> [Accessed 1 April 2015].

Davison, A., 2013. *The Love of Wisdom: An Introduction to Philosophy for Theologians*. London: SCM Press.

Dembski, W.A., 1999. *Intelligent Design: The Bridge Between Science and Theology*. Downers Grove: InterVarsity Press.

Dembski, W.A., 2002. *The Logical Underpinnings of Intelligent Design*. Waco: Baylor University.

Derkse, W., 1993. *On Simplicity and Elegance: An Essay in Intellectual History*. Delft: Eburon.

Derkse, W., 1997. Nice Work: Beauty and Transcendence as Factors in Scientific Practice. In: N. H. Gregersen, M. W.S. Parsons and C. Wassermann, eds. 1998. *The Concept of Nature in Science and Theology (Part II)*. Genève: Edition Labor et Fides. pp. 47–56.

Derkse, W., 2001. One World: The Unwritten Second Part of Wittgenstein's Tractatus. In: W. Desmond, J. Steffen and K. Decoster, eds. 2001. *Beyond Conflict and Reduction: Between Philosophy, Science and Religion*. Leuven: Leuven University Press. pp. 159–173.

DeWolf, L.H., 1958. A Theological Evaluation of Natural Theology. *The London Quarterly and Holborn Review*, 1958 (05). Available at: <https://oxford-institute.org/1958-first-institute/> [Accessed 20 July 2016].

Drees, W.B., 2002. Playing God? Yes! Religion in the Light of Technology. *Zygon*, 37(3), pp. 643–654.

Farley, E., 2001. *Faith and Beauty: A Theological Aesthetic*. Aldershot: Ashgate.

Fodor, J., 2008. 'Alien Beauty': Parabolic Judgment and the Witness of Faith. In: O.V. Bychkov and J. Fodor, eds. 2008. *Theological Aesthetics after Von Balthasar*. Aldershot: Ashgate. pp. 187–200.

Forrest, P., 2013. The Epistemology of Religion. *The Stanford Encyclopedia of Philosophy* (Fall 2013 Edition), E. N. Zalta (ed.), forthcoming URL = <http://plato.stanford.edu/archives/fall2013/entries/religion-epistemology/> [Accessed 31 October 2016].

Foster, M.B., 1936. Christian Theology and Modern Science of Nature (II.). *Mind*, XLV (177), pp. 1–27.

Foster, M.B., 1957. *Mystery and Philosophy*. London: SCM Press.

Fraenkel, P., 1946. *Natural Theology: Comprising "Nature and Grace" by Professor Dr. Emil Brunner and the reply "No!" by Dr. Karl Barth*. London: The Centenary Press.

Fumerton, R., 2006. *Epistemology*. Oxford: Blackwell Publishing Ltd.

Garner, R., 2011. *On Being Saved: the Roots of Redemption*. London: Darton, Longman and Todd.

Geivett, R.D. and Sweetman, B. eds., 1992. *Contemporary Perspectives on Religious Epistemology*. Oxford: OUP.

Gifford, 2016. Available at: <http://www.giffordlectures.org/overview/natural-theology >. [Accessed 8 September 2014].

Gore, C., 1891. *Lux Mundi: A Series of Studies in the Religion of the Incarnation*. Cambridge: CUP.

Grenz, S.J. and Franke, J.R., 2007. *Beyond Foundationalism: Shaping Theology in a Postmodern Context*. Louisville: Westminster John Knox Press.

Gunton, C.E., 1978. *Becoming and Being: The Doctrine of God in Charles Hartshorne and Karl Barth*. Oxford: OUP.

Haack, S., 1995. *Evidence and Enquiry: Towards Reconstruction in Epistemology*. Oxford: Blackwell.

Hancock, A.N., Kavanaghab, Y. and Schiesser, C.H., 2014. The kinetics of alkyl radical ring closures at selenium: formation of selenane. Org. Chem. Front., 2014(1), p. 645. Available at: <http://doi.org/10.1039/c4qo00108g> [Accessed 10 January 2015].

Hansen, S.B., 2010. The Later Wittgenstein and the Philosophy of Religion. *Philosophy Compass*, 5(11), pp. 1013–1022.

Hedley, D., 2013. Literature. In: C. Taliaferro, S.V. Harrison and S. Goetz, eds. 2013. *The Routledge Companion to Theism*. Abingdon: Routledge. pp. 577–587.

Hick, J., 1992. The Rationality of Religious Belief. In: R.D. Geivett and B. Sweetman, eds. 1992. *Contemporary Perspectives on Religious Epistemology*. Oxford: OUP. pp. 304–319.

Hoffmann, R., 2003. Thoughts on Aesthetics and Visualization in Chemistry. *HYLE International Journal for Philosophy of Chemistry*, 9 (1), pp. 7–10.

Holder, R.D., 2013. Natural Theology in the Twentieth Century. In: R. Re Manning, ed. 2013. *The Oxford Handbook of Natural Theology*. Oxford: OUP. pp. 118–135.

Ilichev, V.A., Pushkarev, A.P., Rumyantcev, R.V., Yablonskiy, A.N., Balashova, T.V., Fukin, G.K., Grishin, D.F. and Andreev, B.A., 2015. Luminescent properties of 2-mercaptobenzothiazolates of trivalent lanthanides. *Phys. Chem. Chem. Phys.*, 2015(17), pp. 11000–11005. Available at: <http://dx.doi.org/ DOI:10.1039/C4CP05928J> [Accessed 24 March 2015].

Irenaeus, 2015. Against the Heresies.[online]. Available at: <http://www.gnosis.org/library/advh4.htm> [Accessed 12 March 2015].

Jammer, M., 2009. *Concepts of Mass in Contemporary Physics and Philosophy*. Princeton: Princeton University Press.

Jordan, J., 2013. Pragmatic Arguments and Belief in God. *The Stanford Encyclopedia of Philosophy* (Winter 2013 Edition), E. N. Zalta (ed.), URL = <http://plato.stanford.edu/archives/win2013/entries/pragmatic-belief-god/> [Accessed 31 October 2016].

Kant, I., 2000. *Critique of the Power of Judgment*. Paul Guyer (ed.). Translated from German by P., Guyer and E., Matthews. Cambridge: CUP.

Kaufmann, L., Traulsen, N.L., Springer, A., Schröder, H.V., Mäkelä, T., Rissanenb, K. and Schalley, C.A., 2014. Evaluation of multivalency as an organization principle for the efficient synthesis of doubly and triply threaded amide rotaxanes. *Org. Chem. Front.*, 2014(1), p. 521. Available at: <http://doi.org/10.1039/c4qo00077c> [Accessed 10 January 2015].

Kearney, R., 2011. *Anatheism*. New York: Columbia University Press.

Kim, J., 2011. *Reformed Epistemology and the Problem of Religious Diversity*. Eugene: Pickwick Publications.

Koch, E., Takise, R., Studer, A., Yamaguchi, J. and Itami, K., 2015. Ni-Catalyzed [alpha] α-arylation of esters and amides with phenol derivatives. *Chem. Commun.*, 2015(51), pp. 855–857. Available at: <http://doi.org/10.1039/c4cc08426h> [Accessed 15 December 2014].

Lang, P.F. and Smith, B.C., 2015. An equation to calculate internuclear distances of covalent, ionic and metallic lattices. *Phys.Chem.Chem.Phys.*, 2015(17), p. 3355. Available at: <http://doi.org/10.1039/c4cp05135a> [Accessed 23 January 2015].

Laszlo, P., 2003. Foundations of Chemical Aesthetics. *HYLE International Journal for Philosophy of Chemistry*, 9 (1), pp. 11–32.

Levere, T.H., 2001. *Transforming Matter: A History of Chemistry from Alchemy to the Buckyball*. Baltimore: Johns Hopkins University Press.

Li, W., Huanga, J. and Wang, J., 2013. Organocatalytic conjugate addition promoted by multi-hydrogen-bond cooperation: access to chiral 2-amino-3-nitrile-chromenes.

Org. Biomol. Chem., 2013(11), p. 400. Available at: <http://doi.org/10.1039/c2ob27102h> [Accessed 10 January 2015].

Lycan, W.G., Schlesinger, G.N., 1992. You Bet Your Life: Pascal's Wager Defended. In: R. D. Geivett and B. Sweetman, eds. 1992. *Contemporary Perspectives on Religious Epistemology*. Oxford: OUP. pp. 270–282.

McAllister, J.W.,1989. Truth and Beauty in Scientific Reason. *Synthese*, 78 (1989), pp. 25–51.

McGrath, A.E., 2001. *A Scientific Theology. Volume 1: Nature*. Edinburgh: T&T Clark.

McGrath, A.E., 2004. *The Science of God*. London: T&T Clark.

McGrath, A.E., 2008. *The Open Secret: a New Vision for Natural Theology*. Oxford: Blackwell.

MacIntosh, J.J. and Anstey, P., 2010. Robert Boyle. *The Stanford Encyclopedia of Philosophy* (Fall 2010 Edition), E. N. Zalta (ed.), URL = <http://plato.stanford.edu/archives/fall2010/entries/boyle/> [Accessed 31 October 2016].

MacKinnon, D.M., 1963. Moral Objections. In: D.M. Mackinnon, H.A. Williams, A.R. Idler and J.S. Bezant, eds. 1963. *Objections to Christian Belief*. London: Constable & Co. Ltd. pp. 11–34.

Markham, I., 1998. *Truth and the Reality of God: An Essay in Natural Theology*. Edinburgh: T&T Clark Ltd.

Marshall, B.D., 2000. *Trinity and Truth*. Cambridge: CUP.

Martin, J.D., 1995. *Proverbs*. Sheffield: Sheffield Academic Press.

Meister, C., 2013. Twenty-First-Century Intellectual Life. In: C. Taliaferro, S.V. Harrison and S. Goetz, eds. 2013. *The Routledge Companion to Theism*. Abingdon: Routledge. pp. 153–163.

Mellor, E.B., 1999. *Proverbs*. Oxford: the Bible Reading Fellowship.

Morris, T.V., 1992. Pascalian Wagering. In: R. D. Geivett and B. Sweetman, eds. 1992. *Contemporary Perspectives on Religious Epistemology*. Oxford: OUP. pp. 257–269.

Murphy, N. and McClendon, J. W. Jr., 1989. Distinguishing Modern and Postmodern Theologies. *Modern Theology*, 5(3), pp. 191–214.

Nguyen, M.T., Hung, T.P., Long, V.D., My, P.P. and Minh, T.N., 2015. Fullerene-like boron clusters stabilized by an endohedrally doped iron atom: BnFe with n = 14, 16, 18 and 20. *Phys.Chem.Chem.Phys.*, 2015(17), p. 3000. Available at: <http://doi.org/10.1039/c4cp04279d> [Accessed 23 January 2015].

Nielsen, K., 1973. The Challenge of Wittgenstein: An Examination of his Picture of Religious Belief. *Studies in Religion*, 3(1), pp. 29–46.

Oberheim, E. and Hoyningen-Huene, P., 2013. The Incommensurability of Scientific Theories. *The Stanford Encyclopedia of Philosophy* (Spring 2013 Edition), E. N. Zalta (ed.), URL = <http://plato.stanford.edu/archives/spr2013/entries/incommensurability/> [Accessed 31 October 2016].

O'Donovan, O., 1986. *Resurrection and Moral Order: An Outline for Evangelical Ethics.* Leicester: Inter-Varsity Press.

O'Hanlon, G.F., 1990. *The Immutability of God in the Theology of Hans Urs von Balthasar.* Cambridge: CUP.

Padgett, A.G., 2012. Practical Objectivity: Keeping Natural Science Natural. In: J.B. Stump and A.G. Padgett, eds. 2012. *The Blackwell Companion to Science and Christianity.* Chichester: Wiley-Blackwell. pp. 93–102.

Paley, W., 1881. *Natural History.* New York: American Tract Society.

Pannenberg, W., 1991. *An Introduction to Systematic Theology.* Grand Rapids: Eerdmans.

Pannenberg, W., 2008. *The Historicity of Nature: Essays on Science and Theology.* West Conshohocken: Templeton Foundation Press.

Peacocke, A., 1993. *Theology for a Scientific Age.* London: SCM Press.

Phillips, D.Z., 1992. Faith, Skepticism, and Religious Understanding. In: R.D. Geivett and B. Sweetman, eds. 1992. *Contemporary Perspectives on Religious Epistemology.* Oxford: OUP. pp. 81–91.

Phillips, D.Z., 1993. *Wittgenstein and Religion (Swansea Studies in Religion).* Basingstoke: The Macmillan Press Ltd.

Plantinga, A., 1992. Is Belief in God Properly Basic? In: R. D. Geivett and B. Sweetman, eds. 1992. *Contemporary Perspectives on Religious Epistemology.* Oxford: OUP. pp. 133–141.

Plantinga, A., 1993. *Warrant: The Current Debate.* New York: OUP.

Plantinga, A., 2000. *Warranted Christian Belief.* New York: OUP.

Plantinga, A., 2007. On 'Proper Basicality'. *Philosophy and Phenomenological Research*, LXXV (3), pp. 612–621.

Plantinga, A., 2008. Against Naturalism. In: A. Plantinga and M. Tooley, 2008. *Knowledge of God.* Oxford: Blackwell Publishing Ltd. pp. 1–69.

Plantinga, A., Wolterstorff, N., 1983. *Faith and Rationality: Reason and Belief in God.* Notre Dame: University of Notre Dame Press.

Polkinghorne, J., 2004. *Science and the Trinity: The Christian Encounter with Reality.* London: SPCK.

Prati, L., Villa, A., Chan-thaw, C.E., Campisi, S., Wang, D., Kubel, C., Kotula, P.G. and Bianchi, C., 2015. AuRu/AC as an effective catalyst for hydrogenation reactions. *Phys. Chem. Chem. Phys.*, 2015(17), pp. 28171–28176. Available at: <http://dx.doi.org/DOI:10.1039/C5CP00632E> [Accessed 24 March 2015].

Pseudo-Dionysius, 1987. *The Complete Works*. Translated from Ancient Greek by C., Luibheid. Mahwah: Paulist Press.

Re Manning, R. ed., 2013. *The Oxford Handbook of Natural Theology*. Oxford: OUP.

Richardson, A., 1953. *Genesis 1–11*. London: SCM Press.

Rong, C., Wuming, Y., Bologna, M.G., de Silva, K., Ma, Z., Finklea, H.O., Petersen, J.L., Minyong Li, M. and Shi, X., 2015. Synthesis and characterization of N-2-aryl-1,2,3-triazole based iridium complexes as photocatalysts with tunable photoredox potential. *Org. Chem. Front., 2015(2), p. 141*. Available at: <http://dx.doi.org/DOI:10.1039/c4qo00281d> [Accessed 1 April 2015].

Rowland, C., 2013. Natural Theology and the Christian Bible. In: R. Re Manning, ed. 2013. *The Oxford Handbook of Natural Theology*. Oxford: OUP. pp. 23–37.

Rosenberg, A., 2005. *Philosophy of Science: A Contemporary Introduction*. 2nd ed. Abingdon: Routledge.

Russell, B., 1961. *History of Western Philosophy*. 2nd ed. London: George Allen & Unwin Ltd.

Russell, B., 2013. A Priori Justification and Knowledge. *The Stanford Encyclopedia of Philosophy* (Summer 2013 Edition), E. N. Zalta (ed.), URL = <http://plato.stanford.edu/archives/sum2013/entries/apriori/> [Accessed 31 October 2016].

Sarot, M., 2008. Christian Fundamentalism as a Reaction to the Enlightenment Illustrated by the Case of Biblical Inerrancy. In: B.E.J.H. Becking, ed. 2011. *Orthodoxy, Liberalism, and Adaptation: Essays on Ways of Worldmaking in Times of Change from Biblical, Historical and Systematic Perspectives* (Proceedings of a symposium held in the the South of the Netherlands and Leuven in 2008). Leiden: Brill. pp. 249–267.

Sartwell, C., 2014. Beauty. *The Stanford Encyclopedia of Philosophy* (Spring 2014 Edition), E. N. Zalta (ed.), URL = <http://plato.stanford.edu/archives/spr2014/entries/beauty/> [Accessed 31 October 2016].

Schaefer, K., 2001. *BERIT OLAM. Studies in Hebrew Narrative & Poetry: Psalms*. Collegeville: The Liturgical Press.

Schreiner, T.R., 2013. *The King in His Beauty: A Biblical Theology of the Old and New Testaments*. Grand Rapids: Baker Academic.

Schummer, J., 2003. Aesthetics of Chemical Products. *HYLE International Journal for Philosophy of Chemistry*, 9(1), pp. 73–104.

Schummer, J., 2015. *Die Chemie als Teufelswerk? 2300 Jahre Chemiekritik* [Power Point presentation]. Chemie und Gesellschaft, Deutsches Museum, Munich, Germany, pp. 15–17.

Scruton, R., 2009. *Beauty*. Oxford: OUP.

Seibt, J., 2013. Process Philosophy. *The Stanford Encyclopedia of Philosophy* (Fall 2013 Edition), E. N. Zalta (ed.), URL = <http://plato.stanford.edu/archives/fall2013/entries/process-philosophy/> [Accessed 31 October 2016].

Sherry, P., 2002. *Spirit and Beauty*. 2nd ed. London: SCM Press.

Shook, J.R., 2010. *The God Debates: a 21st Century Guide for Atheists and Believers*. Chichester: Wiley-Blackwell.

Smith, M., 2014. The Epistemology of Religion. *Analysis Reviews,* 74(1), pp. 135–147.

Soskice, J.M., 2007. *The Kindness of God: Metaphor, Gender, and Religious Language*. Oxford: OUP.

Southern, R.W., 1990. *Saint Anselm: A Portrait in a Landscape*. Cambridge: CUP.

Steane, A., 2014. *Faithful to Science: The Role of Science in Religion*. Oxford: OUP.

Stein, R.L., 2004. Towards a Process Philosophy of Chemistry. *HYLE International Journal for Philosophy of Chemistry*, 10(1), pp. 5–22.

Steup, M., 2012. Epistemology. *The Stanford Encyclopedia of Philosophy* (Winter 2012 Edition), E. N. Zalta (ed.), URL = <http://plato.stanford.edu/archives/win2012/entries/epistemology/> [Accessed 31 October 2016].

Stump, J.B., 2012. Natural Theology after Modernism. In: J.B. Stump and A.G. Padgett, eds. 2012. *The Blackwell Companion to Science and Christianity*. Chichester: Wiley-Blackwell. pp. 140–150.

Sudduth, M., 2009. Revisiting the 'Reformed Objection' to Natural Theology. *European Journal for Philosophy of Religion,* 1(2), pp. 37–62.

Tertullian, 2015. *De Cultu Feminarum*. Available at: <http://www.tertullian.org> [Accessed 30th May 2015].

Tuna, D., Sobolewskib, A.L. and Domckea, W., 2014. Electronically excited states and photochemical reaction mechanisms of b-glucose. *Phys.Chem.Chem. Phys.*, 2014(16), p. 38. Available at: <http://doi.org/10.1039/c3cp52359d> [Accessed 2 April 2014].

Van den Brink, G., 2009. *Philosophy of Science for Theologians: An Introduction*. Frankfurt am Main: Peter Lang.

Van Inwagen, P., 2012. Russell's China Teapot. In: D. Łukasiewicz and R. Pouivet, eds. 2012. *The Right to Believe: Perspectives in Religious Epistemology*. Heusenstamm: ontos verlag. pp. 11–26.

Viladesau, R., 1999. *Theological Aesthetics: God in Imagination, Beauty and Art*. Oxford: OUP.

Viney, D., Process Theism, *The Stanford Encyclopedia of Philosophy* (Spring 2014 Edition), E. N. Zalta (ed.), URL = <http://plato.stanford.edu/archives/spr2014/entries/process-theism/> [Accessed 31 October 2016].

Von Balthasar, H.U., 1982. *The Glory of God: A Theological Aesthetics*. Joseph Fessio and John Riches (eds.). Translated from German by E., Leiva-Merikakis. Edinburgh: T & T Clark.

Walhout, P.K., 2009. The Beautiful and the Sublime in Natural Science. *Zygon*, 44(4). pp. 757–776.

Ward, G., 2003. The Beauty of God. In: J. Milbank, G. Ward and E. Wyschogrod, 2003. *Theological Perspectives on God and Beauty*. Harrisburg: Trinity Press. pp. 35–65.

Weisberg, M., Needham, P. and Hendry, R., 2011. Philosophy of Chemistry, *The Stanford Encyclopedia of Philosophy* (Winter 2011 Edition), E. N. Zalta (ed.), URL = <http://plato.stanford.edu/archives/win2011/entries/chemistry/> [Accessed 31 October 2016].

Westermann, C., 1997. Beauty in the Hebrew Bible. In: A. Brenner and C. Fontaine, eds. 1997. *A Feminist Companion to Reading the Bible: Approaches, Methods and Strategies*. Translated from German by U. Östringer and C.R. Fontaine. Sheffield: Sheffield Academic Press. pp. 584–682.

Whitehead, A.N., 1978. *Process and Reality: An Essay in Cosmology*. Corrected Edition. New York: The Free Press.

Wilson, E.O., 1998. *Consilience: The Unity of Knowledge*. London: Abacus.

Wissink, J.B.M., 1993. *Te Mooi Om Onwaar Te Zijn*. Vught: Radboudstichting.

Wittgenstein, L., 1958. *Philosophical Investigations (trans: Anscombe)*. Oxford: Blackwell.

Woolford, T.A., 2011. *Natural Theology and Natural Philosophy in the Late Renaissance*. PhD. Cambridge University. Available at: <http://www.dspace.cam.ac.uk/handle/1810/242394>. [Accessed 8 November 2015].

Wynn, M., 1999. *God and Goodness: A Natural Theological Perspective*. London: Routledge.

Wyschogrod, E., 2003. In: J. Milbank, G. Ward and E. Wyschogrod, 2003. *Theological Perspectives on God and Beauty*. Harrisburg: Trinity Press. pp. 66–86.

Yu, Y., Shu, C., Zhou, B., Li, J., Zhoua, J. and Ye, L., 2015. Efficient and practical synthesis of enantioenriched 2,3-dihydropyrroles through gold-catalyzed anti-Markovnikov hydroamination of chiral homopropargyl sulfonamides. *Chem. Commun.*, 2015(51), p. 2126. Available at: <http://doi.org/10.1039/c4cc09245g> [Accessed 22 January 2015].

Zurek, E. and Wojciech, G., 2015. Predicting crystal structures and properties of matter under extreme conditions via quantum mechanics: the pressure is on. *Phys.Chem.Chem.Phys.*, 2015(17), p. 2917. Available at: <http://dx.doi.org/DOI:10.1039/c4cp04445b> [Accessed 23 January 2015].

Unless otherwise noted all bible texts are quoted from the New English Translation (NET): "Scripture quoted by permission. All scripture quotations, unless otherwise indicated, are taken from the NET Bible copyright (C)1996–2006 by Biblical Studies Press, L.L.C. www.bible.org All rights reserved. This material is available in its entirety as a free download or online web use at http://www.netbible.org/."

Unless stated otherwise diagrams of chemical compounds are taken from http://www.chemspider.com.

Words in Hebrew taken from The Blue Letter Bible CD. CD-ROM, version 2.2. Sowing Circle, 2009. See also: https://www.blueletterbible.org.

The Old Testament Hebrew lexicon is the Brown, Driver, Briggs, Gesenius Lexicon which is available online at http://www.biblestudytools.com/lexicons/hebrew/

Appendices

Appendix A: Some Notes on Chemical Structures

Several diagrams of chemical structures appear in this book. This short appendix is intended to give the reader an insight into what is being signified in these diagrams.

1. The chemistry in this book is restricted to organic chemistry, a sub-discipline concerning those substances built-up largely from carbon. This appendix only relates to such carbon based structures and their visualisation.
2. As will become apparent to the reader, chemistry and the visualisation of its reactants is best comprehended through the use of multiple points of view, which may include mathematical equations, models and diagrams offering a variety of information at differing levels of detail. In this way, the same physical chemical compound (the structure of which obviously is unseen) may be represented by and through a set of diagrams each of which presents not only a different image of what the compound might look like if it could be seen, but also varying levels of efficacy in predicting and illustrating the outcome of reactions in which it may take part.
3. For the purposes of this illustration, an atom is a single physical entity representing an individual instance of an element. The element is therefore the class name for a group of identical physical objects. The term 'element' is familiar to many and includes for example: gold or silver, carbon, oxygen, hydrogen, argon etc. There are a little over 100 of such elements and all chemical compounds are composed of two or more of these in combination. Elements exhibit properties and as a result of segregating or grouping these properties, it has become apparent that there are groups or series of such elements. The well-known 'Periodic Table' is an attempt at classifying elements into groups of similar elements. Most organic compounds are composed of a very much restricted group of elements, typically carbon itself, in combination with hydrogen, sulphur, nitrogen; the so-called halides including chlorine, bromine, iodine and fluorine; certain lighter metals including sodium, lithium, boron and aluminium and then certain heavier metals which form peculiar not-to-say exotic bonding models with the other non-metallic elements listed here and such metals include iron, chromium, copper and even heavier elements such a platinum.

4. Atoms within chemical compounds, and carbon is no exception, become bound to other atoms through the process and phenomena of bonding. There are roughly two categories of bonding (although that statement is fraught with dangers) and for the organic compounds we are considering we should restrict ourselves largely to that class of bonding known as 'covalent'. This involves a process usually characterised as the sharing of one or more electrons that were originally bound to a single atom, such that they would appear to wrap themselves around the atoms within the bond. It is therefore less of an attraction-type 'glue' and should be seen more like ropes. Covalent bonds vary depending on the numbers of electrons being shared and the atoms involved in the bond. This then affects the distance between atomic centres, which itself has a bearing on the nature and energy of the bonding interaction.
5. Organic compounds, meaning those with a 'backbone' composed of one or more carbon atoms, have 'shapes' that are rarely planar in the sense of flat like a sheet of paper. Depending on the type of (covalent) bonding within the compound – and the one compound may have differing types of bonding in different parts of the molecule – it becomes possible for the same collection of individual atoms to be arranged within a compound in possibly many different ways. The resulting set or group of compounds may in turn exhibit remarkably different properties: they may differ in smell, colour, state (i.e. whether gaseous, liquid or solid at a given temperature and pressure), toxicity etc.
6. Bonding between carbon atoms is frequently described as being single, double or triple, referring to the numbers of electrons being shared in the construction of the bond. It is interesting to note that a string of carbons linked by single bonds is shaped somewhat like a three-dimensional saw-tooth, whereas double and triple bonds are planar i.e. flat. This has important implications for the way in which the properties of a compound are expressed in the physical world.
7. Readers will be familiar with the concept of physical structures that are apparently the same, yet cannot be superimposed upon one another. The most obvious being our hands: our two hands are plainly the same and yet cannot be superimposed upon one another. Chemical compounds may exhibit this 'handedness' in a variety of complex ways, such that for example one version of a compound is a useful medicine and the other a largely worthless chemical – yet both would appear to have the same atoms joined together in the same pattern, yet are mirror images of each other. It is this 'stereoscopic' effect, this 'handedness', that is used throughout the chemistry of living things

to construct an enormous variety of 'lock and key' pairs of complex proteins wherever a receptor 'lock' in the body requires a chemical 'key' to open it, for example in order to record a particular stimulus.

8. Colour as a phenomena expressed by chemical compounds occurs because of the manner in which light interacts with that compound. It should be remembered that a compound may exist in a variety of forms. Frequently white light is modified in its passing through for example a solution of a compound, through the removal of certain component wavelengths. At other times light is modified through its being transformed as it hits the surface of a compound. Carbon compounds containing triple carbon bonds are particularly known for often exhibiting colour as are carbon compounds containing metallic atoms.

9. The phenomena of isotopes is already well known in general where a particular element is known in a variety of ways through variations in numbers of constituent particles within the atoms of that element. Thus an element may engage in a particular chemical reaction on account of being of that element and yet the resulting compound may exhibit differing properties because the precise isotope of the element differs. Perhaps a particular well known example is deuterium being an isotope of hydrogen. Being a 'type' or an isotope of hydrogen, it will form water when reacting with oxygen, known popularly as 'heavy water'. If it should however be ingested in any quantity it is a poison since the bonding that 'heavy water' enters into is subtly different to that formed by 'ordinary' water, to the extent that e.g. enzymes in the human body fail to react as they should.

10. Thus variety within chemical compounds may be expressed in a great number of ways: through combinations of atoms, their individual arrangements in three-dimensional space, the isotopes being used, the types of bonding as well as the local context the chemical finds itself in. The science of for example Magnetic Resonance Imaging now routinely used in medical science to see into the human body in a completely non-invasive manner, relies on a particular property of the hydrogen within common water to alter very slightly depending on the local context. In another field, many – often successful – attempts have been made to construct chemical compounds purely with the aim of achieving novel shapes such as for example footballs, cages, cylinders or crowns.

11. A word on notation: a carbon carbon single bond may be written thus C-C; a double bond with two lines between: C=C, and a triple with three. Please remember that these simple binary structures exhibit different arrangements

or structures in physical space. A structure written for example thus C-N represents a single bond between a carbon atom and a nitrogen; C-S would be involving sulphur. At times bonding rather than be shown with a solid single line between atoms, might be displayed with a dotted line. This is particularly the case where the bond can not be easily understood to be a 'simple' single or double bond, but where the electrons involved within the bond may not be as tightly localised. This is particularly the case in bonds involving metallic atoms but also in circumstances where the physical structure is unusual and might allow electrons to be perceived as being more dispersed. Benzene is a popular example: a ring of six carbon atoms would it appears, allow the electrons to be understood as a dispersed 'cloud' surrounding the entire molecule. It is a fascinating compound exhibiting neither entirely single nor entirely double bonds but something in between.

12. It will have become clear to the reader that chemical research is an area of both great variety as well as great complexity. As knowledge has grown, the behaviours of reactants within individual reactions are frequently well understood yet these are then combined in novel ways to produce the astonishing variety of compounds known today, a variety that is expanding constantly. Thus there is 'order' in the sense of known rules and circumstances being used reproduceably. Yet the targets, as in new materials, may be elusive and a number of paths to them may have to be tried before a successful reaction path is identified. Thus the human is making use of predetermined patterns but taking control of them in novel ways, and there is at times great unpredictability in terms of all the parameters of the outcome.

Appendix B: A Brief Introduction to Redox Reactions

Anecdotally it might be expected that in a chemical reaction between several *input* compounds, several *outputs* or *products* might result: the reaction starts with a stable set of compounds, runs to completion and yields another stable set of products. Frequently however this is not the case. Certain sorts of so-called *Redox* reactions, the name being a conflation of the words oxidation and reduction, consume quantities of inputs and then depending on various factors, may or may not run to completion. At a certain point both input compounds and product may co-exist: the reaction can be run forwards and backwards by altering the reaction conditions such as temperature, concentration of reactants and gaseous pressure.

This might be thought a distinct disadvantage however it transpires that redox reactions play a pivotal rôle in many life processes. One of these for example is mammalian respiration. The process by which oxygen is used for respiration is tuneable through the acidity – known in chemistry as the pH – of the blood. The reader may wish to reflect on the behaviour of their own bodies when once they have engaged in exercise: the additional carbon-dioxide in the blood causes the blood to become more acidic which in turn triggers a series of reactions designed to expel the excess carbon dioxide and replenish the blood with oxygen. The blood then returns to a less acidic state: the process overall is reversible it is true but indeed at the level of the individual chemical reactions in the blood, these also are reversible. Plainly the bodily organism is capable of *adaptation* to environmental and other influences in short timespans.

Each redox process involves a movement of electrons with one part of the reaction loosing them and so becoming *oxidised* and another gaining them and so becoming *reduced*. Environmental factors directly influence this flow across the very many groups that might be involved in any redox reaction. Such groups might be relatively simple and consist of single ions or electrically charged metal atoms, or they might consist of large bonded structures consisting of many atoms arranged into very considerably sized, yet still electrically charged, groups. Within such large structures, the arrangement and types of these electrically charged groups or ions, make an enormous difference to how they behave in redox reactions. Elsewhere I have spoken of energy maxima and minima using the example of a ball bearing rolling about a surface. It might be helpful to think of redox systems also moving to a minimum energy consistent with the particular set of environmental conditions. It is possible to draw-up tables predicting how pairs of ions will behave in a redox fashion should they 'meet' each other in such a reaction. Thus for example a particular charged group might habitually react in a particular manner when faced with another reactant. However if this reactant is altered to a 'weaker' or 'stronger' one, the redox reaction may not commence at all. It will rapidly become clear to the reader that reacting systems can be designed to be 'tuneable' depending upon what differing arrangements of atoms are grouped into a basic process-flow or structure of a reaction design. One could envisage for example in the mammalian body, a slight alteration to say an oxygen receiving compound, rendering certain enzyme systems ineffective.

From the perspective of this book, the very design of such systems, most especially those related to life-chemistry but often even ones we might design ourselves, could plainly be seen as elegant and beautiful.

Index of Names

Amesbury 15, 16
Anstey 86, 87
Aquinas 18, 20, 146–148, 150, 155
Aristotle 146–148, 155
Augustine 151, 152
Ayer 146, 149
Bagger 78
Baker 20, 23, 29, 30
Barbour 79, 80, 94, 97, 98, 104
Barr 37, 38
Barrett 46
Barrow 58, 59
Barth 40, 63, 75, 146, 149–154
Bengoetxea 91–93
Bennett-Hunter 68
Berlin 61
Bhaskar 94–96, 103
Boyle 86–88
Brettler 61
Brooke 5, 31, 32, 40, 47, 85, 86, 92, 93, 106, 109, 111, 117, 133
Brümmer 18, 19, 47, 51, 60, 116
Brunner 41, 75
Bulkeley 59, 155
Burnett 75
Bychkov 147, 148
Cantor 31, 32, 40, 47, 85, 86, 92, 93, 106, 109, 111, 117, 133
Cargile 27
Cartwright 90, 91
Casserley 11, 37, 67, 69, 70
Chapman 87
Collins 60
Corrigan 152
Dahling-Sander 75
Davison 77
Dembski 57, 68, 70, 79

Derkse 5, 10, 55, 56, 58, 59, 103, 106, 108, 109, 121, 124, 155
DeWolf 67
Drees 5, 47
Farley 122
Fodor 122, 123
Forrest 13, 17, 25
Foster 41–43, 46, 52–54, 87, 124
Fraenkel 75
Franke 19
Fumerton 21
Garner 110, 111
Geivett 18, 23
Gifford 9, 85, 110, 114
Gore 45, 52, 53
Grenz 19
Gunton 75, 115
Haack 93
Hansen 15, 16
Harrington 152
Hedley 78
Hick 28
Hoffman 105
Holder 38, 40
Hoyningen-Huene 48
Irenaeus 101
Jammer 48
Jordan 26, 27
Kant 146, 148, 149
Kearney 146, 148, 149
Kim 20, 24
Lazslo 118
Levere 88, 90, 92, 93
Lujan 91–93
Lycan 27
MacIntosh 86, 87
MacKinnon 32, 33

Markham 114, 115
Marshall 14, 82
Martin 13, 25, 101
McAllister 59
McClendon 68, 69
McGrath 24, 39–41, 44–46, 50, 52, 54, 59, 72, 73, 75, 76, 94–96, 104, 105, 116, 153
Meister 10, 11, 38
Mellor 101
Morris 26
Murphy 68, 69, 101
Needham 88, 89
Nielsen 15
O'Donovan 65, 66
O'Hanlon 154
Oberheim 48
Padgett 105, 106, 108
Paley 56, 57
Pannenberg 25, 40, 48, 49, 52, 68, 119
Peacocke 112
Phillips 15–17
Plantinga 17, 18, 20–24, 29, 30, 60, 78, 93, 125, 164
Plato 146, 147, 151, 154
Polkinghorne 72, 95, 97–99, 103, 104, 112, 116, 117, 159
Pseudo-Dionysius 146, 151, 152
Re Manning 38, 40, 45, 47, 67
Richardson 60
Rosenberg 57, 59
Rowland 66, 67
Russell, Bertrand 68
Russell, Bruce 41
Sarot 5, 18
Sartwell 145
Schaefer 71
Schlesinger 27
Schreiner 62–64
Schummer 5, 86, 107, 108
Scruton 120, 121
Seibt 111, 114
Sherry 60–63, 102
Shook 28, 29, 58
Smith 13, 22, 25, 26, 31, 33–35
Soskice 94, 95, 97, 98, 103
Southern 68
Steane 69, 70
Stein 113
Steup 13, 19, 34
Stump 43
Sudduth 74–76
Sweetman 18, 23
Tertullian 86
Todt 91–93
Van Den Brink 16–18
Van Inwagen 30
Viladesau 145, 146
Viney 110, 111
Von Balthasar 146, 153, 154
Walhout 135, 149
Ward 64, 121, 123
Weisberg 88, 89
Westermann 61, 62, 161
Whitehead 110, 111, 114, 115, 146, 154, 155
Wilson 53, 79
Wissink 63
Wittgenstein 15–17, 95
Wolterstorff 18, 29, 30
Woodford 76
Wynn 11, 21, 24, 64, 69, 70, 73, 82, 157, 158, 160–164
Wyschogrod 120

Index of Subjects

aesthetics, aestheticism
  as lifestyle 122, 159
  categorizing natural theologies 67
  common language in
    science and theology 109
  in Ayer 149
  in Barth 153
  in chemistry 107–109
  in Kant 148
  in natural theologies 58
  in Pseudo-Dionysius 152
  in language of chemistry reports
    130, 132, 135, 136
atheist, atheistic
  objections to natural theology
    78, 79
  militant 160
attitude
regulating perceptive ability 65, 77,
  106, 145, 155
beauty
  and God 145–156
  as bridge 118
  as other 122
    definition 157–162
    in art 119
  in Barth 149, 150
    in scripture 61, 62
  of Christ 63, 64, 73, 155
bible
  and chemistry 117
  attitude towards 10, 99
  in natural theology 60
  and beauty 61
chemistry
  and aesthetics 104–109, 127, 139
  and beauty 105, 108

  and bonding 177–180
  and order 40
  and process philosophy 113, 114
  and the natural 45–47
  and theology 48, 49, 87
  organic 92
  research papers 127–141
  use of multiple sampling
    methodologies 42, 43
Christ
  and beauty *see beauty of Christ*
  as focus 50
  in Whitehead's philosophy 114
  identity 99, 100
  perceived in creation 52, 98–103
coherentism 19
conversation
  as device used in the natural theol-
    ogy in this book 37, 40, 44, 45,
    66, 81, 82
  as context for this book 10, 11
  as *telos* 123, 160
  certain natural theologies less
    conducive for 69
  Jesus in 153
  In theologies of nature 80
complementarity 158–160, 163, 164
complexity 79
  aspect of beauty 64
  in chemistry 88, 92
  in natural theology 47, 55–58, 67,
    68, 70
  efforts to reduce 108, 116, 118, 164
critical realism 94–98, 103, 104, 164
dualism 50, 52, 53
epistemology 9, 11, 69, 82
  religious 13–35

evidence 13, 21, 23, 33–35
　for personal experience of God 29, 30
　for possible indication of design 57
　nature of 24–26
fideism
　in critical realism 95
　Wittgensteinan 15–17
foundationalism 14, 18, 19, 53, 68, 70
incommensurability 58
　resolving between chemistry and theology 48, 49
justification 13, 14, 16, 17, 25, 31, 34, 35
　and reliabilism 26
　and simplicity 56
　attitude conducive to 77
　in Plantinga 20–24
　of belief by prudential accounts 26, 27, 44
　of belief by religious experience 27–30
models, modelling
　in chemistry 91, 93, 97, 105, 122, 124, 129–133, 135–138, 143
　in critical realism 104, 159
　in simplification 118, 164
　in theology 97, 98
　together in science and theology 108

nature
　and the natural 44–48, 53
　theology of 67, 79–81
order 40–44
process philosophy 110, 111
Reformed Epistemology 9, 17, 19–24
reliabilism 26, 34, 82, 93
religious experience 27, 29–31, 35, 82
similarity
　as bridge between disciplines 67, 70, 107
　as investigative tool 91, 92, 118, 127, 142
simplicity 42, 55–58, 64, 68, 88, 108, 116, 164
scripture 10, 17, 21, 27, 28, 32, 45, 47, 51, 54
　and process philosophy 112
　and the trinity 99–103
　in natural theology 60–67
　showing God as beautiful 60–64, 123
theology
　natural 10, 37–83
　of nature *see nature, theology of*
transcendence
　in beauty 121
　in natural theology 50–55
　in process philosophy 112
trinity *see scripture, and the trinity*
warrant 29, 30, 44
　in epistemology 21, 22, 26–28, 91–93, 106, 125, 164

Table of Figures

Figure 1. Benzene 89

Figure 2. 1,4- and 1,2-dibromobenzene 90

Figure 3. Fullerene 134

Figure 4. Indene 140

Figure 5. Andirolactone 140

Contributions to Philosophical Theology

Edited by Gijsbert van den Brink, Joshua R. Furnal and Marcel Sarot

Vol. 1 Gijsbert van den Brink / Marcel Sarot (eds.): Understanding the Attributes of God. 1999.

Vol. 2 Marcel Sarot / Gijsbert van den Brink (eds.): Identity and Change in the Christian Tradition. 1999.

Vol. 3 Marcel Sarot: Living a Good Life of Evil. 1999.

Vol. 4 William Hasker / David Basinger / Eef Dekker (eds.): Middle Knowledge. Theory and Applications. 2000.

Vol. 5 Wybren de Jong: Identities of Christian Traditions. An Alternative for Essentialism. 2000.

Vol. 6 Eeva Martikainen (ed.): Infinity, Causality and Determinism. Cosmological Enterprises and their Preconditions. 2002.

Vol. 7 Gerrit Brand: Speaking of a Fabulous Ghost. In Search of Theological Criteria, with Special Reference to the Debate on Salvation in African Christian Theology. 2002.

Vol. 8 Guus Labooy: Freedom and Dispositions. Two Main Concepts in Theology and Biological Psychiatry, a systematic Analysis. 2002.

Vol. 9 Wilko van Holten: Explanation within the Bounds of Religion. 2003.

Vol. 10 Santiago Sia: Religion, Reason and God. Essays in the Philosophies of Charles Hartshorne and A.N. Whitehead. 2004.

Vol. 11 Arjan Markus: Beyond Finitude. God's Transcendence and the Meaning of Life. 2004.

Vol. 12 Gijsbert van den Brink: Philosophy of Science for Theologians. An Introduction. 2009.

Vol. 13 Sze Sze Chiew: Middle Knowledge and Biblical Interpretation. Luis de Molina, Herman Bavinck, and William Lane Craig. 2016.

Vol. 14 Timothy Weatherstone: Reconstructing Wonder. Chemistry Informing a Natural Theology. 2017.

www.peterlang.de